"双一流"建设精品出版工程

Physics Laboratory Manual

Editors in chief: Junqing Li Guangyu Fang
Subeditors: Shan Lin Jinwei Gao Weilong Liu Ancai Wu

哈爾濱工業大學出版社
HARBIN INSTITUTE OF TECHNOLOGY PRESS

Introduction

The purpose of this book is to help students to understand the theoretical knowledge from university physics course, meanwhile to cultivate their scientific spirit and quality, and enhance their practical ability to analyze and solve problems. This book is divided into three sections. In the section of introduction, he significance of physical experiments, the rules and skills for data processing are emphasized; in the section of experimental topics, more than 30 topics are designed covering mechanics, heat, acoustics, optics, electromagnetics and modern physics; the readings section gives the basic physical constants, the Nobel Laureates in physics and reasons, and details of some of the most beautiful physics experiments in human history.

This book is suitable for the science and engineering undergraduates (including international students and students who want to study abroad) and the teachers who want to teach and arrange teaching this course.

图书在版编目(CIP)数据

物理实验讲义 = Physics Laboratory Manual:英文/李俊庆,方光宇主编.—哈尔滨:哈尔滨工业大学出版社,2021.9

ISBN 978-7-5603-9168-7

Ⅰ.①物… Ⅱ.①李… ②方… Ⅲ.①物理学-实验-高等学校-教材-英文 Ⅳ.①O4-33

中国版本图书馆 CIP 数据核字(2020)第 222006 号

策划编辑	王桂芝
责任编辑	周一瞳 李青晏
出版发行	哈尔滨工业大学出版社
社　　址	哈尔滨市南岗区复华四道街 10 号　邮编 150006
传　　真	0451-86414749
网　　址	http://hitpress.hit.edu.cn
印　　刷	黑龙江艺德印刷有限责任公司
开　　本	787mm×1092mm　1/16　印张 21.25　字数 544 千字
版　　次	2021 年 9 月第 1 版　2021 年 9 月第 1 次印刷
书　　号	ISBN 978-7-5603-9168-7
定　　价	59.80 元

(如因印装质量问题影响阅读,我社负责调换)

Preface

Physics belongs to the category of science, which studies the fundamental laws of the motion in variety of forms.

Physics was branched from the philosophy originally but it exerts the persistent influence on the other branches of science like chemistry, biology, etc. The concepts, ideas or thoughts, the methods and invented tools in physics promote and accelerate the development of science and technology. Physics plays an important part in the contemporary society.

Physics, anyhow, is based on facts and practice. It needs validation and check from the experiments. Any theory, no matter how perfect and impeccable it seems, needs to be tested by practice.

Wilhelm Conrad Röntgen, the first laureate of Noble prize in physics said, "The experiment is the most powerful and reliable lever enabling us to extract secrets from nature. The experiment must constitute the final judgment as to whether a hypothesis should be remained or be discarded."

The greatest physicist of our time, Albert Einstein made emphasis many times that "it is sufficient for one contradictory experimental result to overthrow a theory". In fact, the gravitational wave, which was predicted by his general theory of relativity nearly one hundred years ago, has become one of the hot topics in scientific community at once since the issue of the experimental achievement from LIGO (Laser Interferometer Gravitation-wave Observatory) and the photo of black hole from the EHT (Event Horizon Telescope) recent years.

From the viewpoint above, one may feel the significance of the experiments, especially in physics.

"Physics laboratory" or "physics experiments", as a relatively independent course, may augment and supplement the knowledge learnt in college physics by students, providing the comprehensive training in many aspects, not only in the experimental skills and methods but also in the scientific attitude and the scientific spirit for the students who may be engaged in scientific or engineering works now or in future. We hope the course may exert active influence on the students in self-improvement and individual development.

Concretely, the main purpose and the aim of the course lies in the following aspects:

(1) To reinforce the understanding of physics principles learnt from the theoretical courses;

(2) To provide a basic training in the scientific experimental methods and skills;

(3) To train the ability to independently or solely carry on the scientific experimental research;

(4) To cultivate the scientific quality such as the attitude of objectivity, earnest, being practical and realistic, the courage to face the difficulties and to exploit and reveal the truth of the facts, the spirit of team and cooperation.

We hope this book—*Physics Laboratory Manual*, will provide the students who need the training in *physics experiments the supporting materials*, and provide an assistance to the teachers who have a willing to arrange the course and guide the students to perform the experiments successively.

The authors owe a debt of gratitude to many colleagues who have contributed to this book. Special thanks are to Professor Haifa Zhao who provided us the firm support and valuable suggestions, and to Associated Professor Li Huang who contributed her manuscript on the Franck-Hertz experiment. We are grateful to Associated Professor Qingxin Yang for giving us valuable suggestion in organization of this manual. Also special thanks to Professor Jianlong Liu who is working at Harbin Engineering University (HEU) now, to Senior Engineer Dakun Wu and Lecturer Meng Tang who supported us in arranging the experiments and teaching the foreign students.

Junqing Li, Guangyu Fang
March, 2021
Harbin Institute of Technology

Contents(目录)

Part 1(第1部分) Introduction to Laboratory Physics—About Measurement Errors, Data Treatment and Analysis(实验物理绪论——关于测量误差、数据处理及分析) 1

  1.1 Measurement and Errors(测量与误差) 1

  1.2 Classification of Errors(误差的分类) 2

  1.3 Treatment of Random Errors(随机误差的处理) 5

  1.4 Treatment of Systematic Errors(系统误差的处理) 9

  1.5 Uncertainty of Measured Results(测量结果的不确定度) 9

  1.6 Expression of Direct Measured Results(直接测量结果的表达式) 10

  1.7 Expression of Indirect Measured Results(间接测量结果的表达式) 11

  1.8 Significant Figures (Digits) and Their Algorithm(有效数字及其计算) 14

  1.9 Treatment of Experimental Data—Arrangement, Demonstration and Analysis(实验数据处理——整理、图示和分析) 18

  1.10 Exercises(练习题) 22

Part 2(第2部分) Topics of Experiments(实验项目) 24

  2.1 Measurement of Density(密度测定) 24

  2.2 Measurement of Young's Elastic Modulus(杨氏弹性模量的测定) 31

  2.3 Measurement of Liquid's Viscosity(液体黏度的测定) 37

  2.4 Measurement of the Rotational Inertia of a Rigid Body(刚体的转动惯量测定) 44

  2.5 Collision-Shooting Experiment(射击打靶) 51

  2.6 Standing Waves on a String(弦上驻波) 57

  2.7 Exploring the Musical Sounds(研究乐音) 64

  2.8 Forced Vibration of Pohl's Pendulum(波尔摆的受迫振动) 82

  2.9 Thermocouple and Its Application(热电偶及其使用) 91

  2.10 Wheatstone Electrical Bridge and Its Application(惠斯通电桥及其应用) 98

  2.11 Measuring the Temperature Coefficient of Resistance Using Unbalanced Bridge(用非平衡电桥测量电阻的温度系数) 103

  2.12 Potentiometer and Its Application(电位差计及其应用) 109

  2.13 Temperature Characteristics of PN Junction(PN结的温度特性) 115

  2.14 Oscilloscope and Its Applications(示波器及其应用) 123

  2.15 Measurement of Sound Velocity in Air(空气声速的测定) 137

  2.16 Voltmeter-Ammeter Method for Measurement of Resistance(伏安法测量电阻) 144

  2.17 Transient Process of a RC Circuit(RC电路的瞬态过程) 150

  2.18 Measuring Magnetic Fields of Helmholtz Coil(亥姆霍兹线圈磁场测量) 158

  2.19 Dynamic Hysteresis Loop of Ferromagnetic Material(铁磁材料的动态磁滞回线) 167

2.20 Hall Effect and Its Application in Measuring the Magnetic Field(霍尔效应及其在磁场测量方面的应用) …………………………………………………… 176

2.21 Measuring the Focal Length of Thin Lens(测定薄透镜焦距) ………… 183

2.22 Self-assembling a Microscope and a Telescope(组装显微镜和望远镜) ……… 192

2.23 Measuring the Wavelengths of Light with a Transmission Diffraction Grating(透射光栅测量光波长) ……………………………………………………… 205

2.24 Study Material Dispersion of a Prism(研究棱镜材料的色散关系) ………… 213

2.25 Equal Thickness Interference and Its Applications(等厚干涉及其应用) ……… 220

2.26 Polarized Light: Generation and Determination(偏振光:产生与测定) ……… 227

2.27 Holography Technique(全息照相术) …………………………………… 236

2.28 Michelson Interferometer(迈克尔孙干涉仪) …………………………… 244

2.29 Frank-Hertz Experiment(弗兰克-赫兹实验) …………………………… 254

2.30 Determining the Charge of an Electron by Millikan's Method(密立根油滴测定电子电荷) …………………………………………………………… 265

2.31 Optical Fiber Transmission Technique(光纤传输技术) ………………… 271

Part 3(第 3 部分) Reading Materials(阅读材料) ……………………… 283

3.1 The Fundamental Physical Constants(物理学基本常数表) …………… 283

3.2 List of Laureates of Noble Prize in Physics(物理学诺贝尔奖获得者) ……… 284

3.3 List of 10 Most Beautiful Experiments in Physics (10 个最美丽的物理实验) ……… 303

3.4 More About 10 Most Beautiful Physics Experiments(更多关于 10 个最美丽的物理实验) …………………………………………………………… 304

References(参考文献) ……………………………………………………… 332

Part 1 Introduction to Laboratory Physics —About Measurement Errors, Data Treatment and Analysis

1.1 Measurement and Errors

In the practical activity of human being's daily life and work, especially in science and engineering, there is often a need to get the quantitative information on an objective or body being observed and measured. The process to get the information is called the <u>measurement</u>. In fact, the measurement is a process to compare the unknown to the known.

Generally, the measurement may be classified into <u>two types: direct one and indirect (intermediate) one</u>. For instance, the length (extent) of an object can be obtained by using a simple ruler or more complex instrument such as the vernier caliper and micrometer (microcaliper). It can be viewed as a <u>direct measured quantity</u> while the volume of a box as an <u>indirect measured quantity</u> because to get it one firstly needs to get the all extents such as length, width, height (the direct measured quantities) of the box (volume = length × width × height, $V = l \times w \times h$). So the volume V is an indirect measured quantity. An indirect measured quantity is always linked to some direct measured quantities via a certain relation such as a mathematical function.

The measurement is often taken under some condition or environment: the imperfection of the instruments; the fluctuation in the measurement condition and environment (such as the fluctuation of temperature, mechanical vibration); nonstandard operation of the experimenter may result in the <u>errors of measurement</u>; the deviation, departure or divergence from the <u>true value</u>. It makes the measurable quantity or value with some <u>uncertainty</u> after a measurement process. Note that the true value is the real value under a given objective condition, as it is pointed out later that the existence of errors is inevitable, so it is difficult to accurately catch it but one can approach it.

Usually, in order to evaluate the measured quality one may compare the <u>measured quantity</u> a with the <u>true value</u> a_0, the difference between them is called the <u>absolute error (absolute difference)</u> Δa:

$$\Delta a = a - a_0 \qquad (1.1.1)$$

Only the absolute error, however, could not reflect the accuracy of the measured result, so we need another evaluation standard (criterion), the <u>fractional error (relative error or</u>

percentage difference) E:

$$E = \frac{\Delta a}{a_0} \times 100\% \qquad (1.1.2)$$

1.2 Classification of Errors

According to the reasons how the errors arise, the degree to them mastered or known by the experimenter and their properties or the laws that obey, one can classify the errors into three categories.

1. Systematic (determinate) errors

Sometimes, taking repeated measurement of a quantity under "the same condition", we find that the magnitude and the sign of all the errors are kept unchanged (constant), or obeys a certain law. We call such a kind of error the systematic error or determinate error. Systematic errors may result from the imperfection of the instruments or the methods and technique used in the experiments, or from the inappropriate use of the instruments or from the biased, bad habit or physiological defect or bad psychology of an experimenter — called as individual (personal) error. The term systematic implies that same magnitude and sign of experimental uncertainty are obtained when the measurement is repeated. Determinate means that the magnitude and sign of the uncertainty can be determined if the error is identified. Here we give some examples of systematic errors.

(1) An improperly "zeroed" or "uncalibrated" instrument, for example, an ammeter as shown in Fig. 1.2.1.

Fig. 1.2.1　An improperly zeroed instrument—the ammeter, which has no current through it, would systematically give an incorrect reading larger than the true value (After correcting the error by zeroing the meter, which scale would you read when using the ammeter?)

(2) A faulty instrument, such as a thermometer that reads 102 ℃ when immersed in boiling water at standard atmospheric pressure. This thermometer is faulty because the reading should be 100 ℃.

(3) Personal error, such as using a wrong constant in calculation or always taking a high or low reading of a scale division. Reading a value from a measurement scale generally involves aligning a mark on the scale. The alignment and hence the value of the reading can depend on the position of the eye (parallax). Examples of such personal systematic error are shown in Fig. 1.2.2.

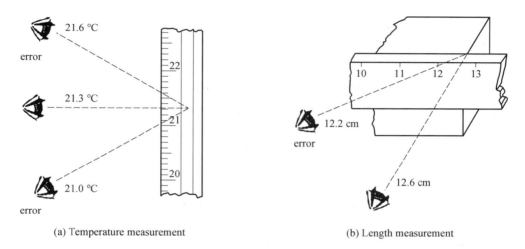

(a) Temperature measurement   (b) Length measurement

Fig. 1.2.2 Examples of personal error due to parallax in reading a thermometer and a meter stick. Readings may systematically be made either too high or too low

(4) Systematic error may arise from the over-simplified formulation of relation in physical amounts in the experiment or lower theoretical approximation. As we know that in the experiment of measuring the acceleration of gravity (g) if using the simple single pendulum, the limitation to the swing angle θ is that it must be less than 5°. When the angle is larger than the limit we should use the series expansion to accurately express the acceleration of gravity:

$$g = \frac{4\pi^2 l}{T^2}\left(1 + \frac{1}{4}\sin^2\frac{\theta}{2} + \frac{9}{64}\sin^4\frac{\theta}{2} + \cdots\right)^2 \qquad (1.2.1)$$

where l and T denote the length of a single pendulum and period of the single swing or time for round trip. One may find that the first term is dominant. When the angle θ is small enough the contribution from the terms related to the angle can be ignored with sufficient-higher accuracy. Thus with a small swing angle we have

$$g = \frac{4\pi^2 l}{T^2} \qquad (1.2.2)$$

which is the expression used to indirectly determine the acceleration of gravity (g) by directly measuring the length (l) of the single pendulum and period (T) of the single swing.

Note that usually one could not find the existence of systematic errors only by repeating measurement in a fixed method under a certain condition unless he or she changes the method or instrument and checks the theoretical approximation degree. In fact, avoiding systematic errors depends on the skill of the observer to recognize the sources of such errors and to prevent or correct them.

2. Random (indeterminate) or statistical errors

Random errors result from unknown and unpredictable variations that arise in all experimental measurement situations. The term indeterminate refers to the fact that there is no way to determine the magnitude or sign (+, too large; −, too small) of the error in any single measurement. Such a kind of error has the following features: it is inevitable, irremovable, random in a single measurement, but obeying some statistical law in repetitious measurement.

Conditions in which random errors can result include:

(1) Unpredictable fluctuations in measurement environment such as temperature or line voltage;

(2) Mechanical vibrations of an experimental setup;

(3) Unbiased estimates of measurement readings by the observer, repeated measurements with random errors give slightly different values each time, the effect of random errors may be reduced and minimized by improving and refining experimental techniques.

3. Coarse errors or mistakes

The coarse (rough or gross) errors are such kinds of errors that obviously depart or deviate from the expectation or break the rule that the most data follow. Coarse errors may result from carelessness or negligence of the observer, or the sudden transition of the experiment condition. For example, by disordered writing the figure 7 may be recorded or recognized as figure 1. Sometimes one can directly cancel such a "poor" datum and repeat the measurement in the same condition after carefully checking the reason.

Repeating the experiment and checking the "strange" results is a good habit that may lead the experimenter to discovery of the new things and laws. Intentionally picking up or selecting the "good" data, however, may be viewed as violating the basic scientific spirit.

All kinds of errors are categorized in the Table 1.2.1.

Table 1.2.1   The classification of errors

By the sources of generation	By the degree of mastering or knowing	By the properties or laws obeyed
◇Error from the instrument ◇Error from the environment ◇Error from the method ◇Error from individual difference	◇Known (determinate) error ◇Unknown (indeterminate) error	◇Systematic error ◇Random error ◇Coarse error

4. The analogy of systematic error and random error with the concepts of accuracy and precision

Accuracy and precision are commonly used synonymously, but in experimental measurements there is an important distinction. The accuracy of a measurement signifies how close it comes to the true (or accepted) value, that is, how nearly correct it is. Precision refers to the agreement among repeated measurements — the "spread" of the measurements or how close they are together. The more precise a group of measurements, the closer together they are. However, a large degree of precision does not necessarily imply accuracy, as illustrated in Fig. 1.2.3. In the practice of shooting, the targets are the bull's eyes. The true value in this analogy corresponds to the center of a target. The degree of scattering is an indication of precision, the closer together a grouping of bullet's impact points, the greater the precision. A group (or symmetric grouping with an average) close to the true value represents the accuracy. Acquiring greater accuracy for an experimental value depends in general on minimizing systematic errors. Acquiring greater precision for an experimental value depends on minimizing random errors.

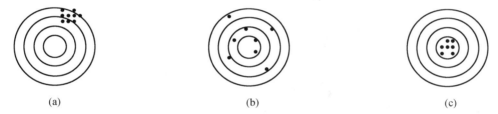

(a)   (b)   (c)

Fig. 1.2.3   Analogue of accuracy and precision, in which (a) corresponds to case with a larger systematic error, (b) to the case with larger random errors, while (c) to the both kinds of errors that are acceptable (adequately small)

1.3 Treatment of Random Errors

Sometimes the data in the repetitive measurement demonstrate a certain statistic law, and the random errors are obvious. In the same condition, we measure a physical quantity A by n times, so we get an array of measurement data:

$$A_1, A_2, A_3, \cdots, A_n$$

If we know the true value A_0, the absolute difference (absolute error) of every measurement $\Delta A_i (i = 1, 2, 3, \cdots, n)$ can be easily obtained:

$$x_i = \Delta A_i = A_i - A_0$$

One can plot a statistic histogram to show the dependence of appearance frequency on the absolute error, as shown in Fig. 1.3.1. It is obvious that the smaller the absolute error, the larger the probability (the more appearance times), and vice versa.

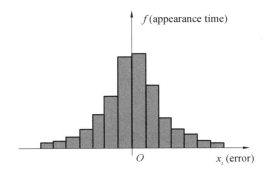

Fig. 1.3.1   The statistic histogram for the measurement with obvious random errors

When the times of measurement approaches infinity (in fact it is not possible to carry the measurement with infinite times), the statistics become continuous, the profile of the histogram can be expressed by the so-called function of normal distribution (Gaussian function):

$$f(x) = \frac{1}{\sqrt{2\pi}\sigma} e^{-\frac{x^2}{2\sigma^2}} \qquad (1.3.1)$$

where σ is termed as the standard error, in fact it is the RMS (root-mean-square) of the errors:

$$\sigma = \sqrt{\frac{1}{n}\sum_{i=1}^{n} x_i^2} = \sqrt{\frac{1}{n}\sum_{i=1}^{n}(A_i - A_0)^2}, \quad n \to \infty \qquad (1.3.2)$$

The plot of normal distribution $f(x)$ is shown in Fig. 1.3.2. $f(x)$ is also called the probability density function. The area enclosed by shadowed zone denotes the probability or possibility that the errors fall into the range from x to $x + dx$. So the probability for random errors appearing in an interval from a to b can be obtained by a definite integral:

$$P(a \leq x \leq b) = \int_a^b f(x) \, dx \qquad (1.3.3)$$

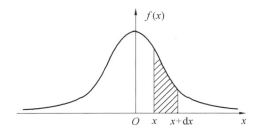

Fig. 1.3.2   The function of normal distribution with a fixed σ

Part 1 Introduction to Laboratory Physics—About Measurement Errors, Data Treatment and Analysis

The range $[a, b]$ and probability $P(a \leqslant x \leqslant b)$ denote the <u>confidence interval</u> and corresponding <u>confidence probability</u> (level). Obviously, the integral for whole range errors must be unity:

$$P(-\infty \leqslant x \leqslant +\infty) = \int_{-\infty}^{\infty} f(x)\,dx = 1 \quad (1.3.4)$$

which means the random errors occur in every case with the probability of 100%. One can easily prove this by performing the complete integration of $f(x)$.

It should be noticed that the function of normal distribution $f(x)$ is a symmetric or even function, which has a maximum $f(0) = \dfrac{1}{\sqrt{2\pi}\,\sigma}$ when $x = 0$ and the two points of inflexion correspond to the positions of $x = \pm\sigma$. It is interesting that

$$P(-\sigma \leqslant x \leqslant +\sigma) = \int_{-\sigma}^{+\sigma} f(x)\,dx = 0.683 \quad (1.3.5)$$

which means that the possibility of random errors located in $[-\sigma, +\sigma]$ is 68.3%. In the same way, we carry on the integration in range $[-2\sigma, +2\sigma]$ and $[-3\sigma, +3\sigma]$ and get the probability of 95.5% and 99.7% (almost unity — the whole area below the curve of $f(x)$, see the Fig. 1.3.3). Hence one can see the role of the standard error.

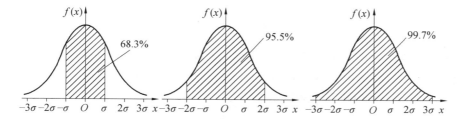

Fig. 1.3.3 The confidence intervals and confidence probabilities (levels) corresponding to $[-\sigma, +\sigma]$, $[-2\sigma, +2\sigma]$ and $[-3\sigma, +3\sigma]$

The value of standard error σ indicates the degree of concentration or the similarity of measured data. Because the whole area covered by the curve is fixed to unity, the larger the standard error, the lower the curve is, vice versa (Fig. 1.3.4). The smaller σ suggests better degree of concentration of data or smaller contribution of random errors. Thus by acquiring the information of the standard error, we may estimate the quality of the measurement. However, it is necessary to mention again that in practice we could not perform the measurement for infinite times and master the true value on the firm sense. The standard error seems to make sense only in theory. To avoid this problem, we introduce some useful concepts about measurement of finite times.

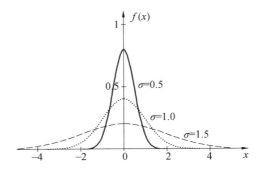

Fig. 1.3.4 The value of standard error σ indicates the degree of concentration of the measurement data or the contribution of the random errors

1. The arithmetic average of data in a series of measurement

In the same condition, measuring the physical quantity A by n times forms a series of measurement: $A_1, A_2, A_3, \cdots, A_n$. The arithmetic average (mean value) is expressed by

$$\bar{A} = \frac{1}{n}\sum_{i=1}^{n} A_i, \quad i = 1, 2, 3, \cdots, n \tag{1.3.6}$$

Due to the compensability of the random errors in the same condition by repetitively measuring a physical quantity, the arithmetic average of error approaches to zero with the increasing measurement times. So we have

$$\bar{A} = \lim_{n \to \infty} \frac{1}{n}\sum_{i=1}^{n} A_i = A_0 \tag{1.3.7}$$

The arithmetic average of limited times $\bar{A}$ is a good approximation of the true value or the best estimate. We call the difference of each measured value of the physical quantity with its arithmetic average the <u>residual error</u>:

$$\nu_i = A_i - \bar{A}, \quad i = 1, 2, 3, \cdots, n \tag{1.3.8}$$

which can be used to evaluate the deviation of each measurement.

2. The standard deviation of a series of limited measurement

<u>The standard deviation of a series of limited measurement</u> is defined as

$$S_A = \sqrt{\frac{1}{n-1}\sum_{i=1}^{n} \nu_i^2} = \sqrt{\frac{1}{n-1}\sum_{i=1}^{n} (A_i - \bar{A})^2} \tag{1.3.9}$$

when the measurement time n approaches to infinity S_A becomes a RMS of the residual errors because there is no difference between n and $n - 1$. The above expression is the famous Bessel's formula which is very useful in practical engineering calculation. As the estimate of the standard error, S_A may be used to evaluate the measured quality.

3. The standard deviation of mean values for limited measurement

In practice one may find that when the whole measurement series is divided into several sub-series (m groups with $m < n$), whose mean value $\overline{A}_j (j = 1,2,3,\cdots,m)$ is different with trivial fluctuation. In fact, $\overline{A}_j$ also follows the normal distribution. So there exists a standard deviation of mean values for sub-series $S_{\overline{A}}$, which is less than the standard deviation of the whole series S_A, $S_{\overline{A}}$ is defined as

$$S_{\overline{A}} = \frac{S_A}{\sqrt{n}} = \sqrt{\frac{1}{n(n-1)} \sum_{i=1}^{n} (A_i - \overline{A})^2} \qquad (1.3.10)$$

It is confirmed that the role of standard deviation of mean values $S_{\overline{A}}$ is similar to the standard error σ as mentioned above or measurement of the limited time: the true value A_0 falls in the interval of $[\overline{A} - S_{\overline{A}}, \overline{A} + S_{\overline{A}}]$ with a probability of 68.3%, in the interval of $[\overline{A} - 2S_{\overline{A}}, \overline{A} + 2S_{\overline{A}}]$ with a probability of 95.5%, in the interval of $[\overline{A} - 3S_{\overline{A}}, \overline{A} + 3S_{\overline{A}}]$ with a probability of 99.7%. Hence, a standard deviation of mean values $S_{\overline{A}}$ may be used to replace the standard error σ to evaluate the quality of measurement of finite times. It is clear that the wider the confidence interval, the higher the confidence probability of A_0 falls in, and vice versa. In the real case, the number of measurement of finite times n takes 5 ~ 10 because with n increase (> 10), $S_{\overline{A}}$ slows down in decreasing, which means measurement of several times is enough for higher accuracy.

1.4 Treatment of Systematic Errors

The systematic errors are mainly classified into two kinds: the determinate ones and indeterminate ones.

For the former the variation law, magnitude and sign are determinate while for the latter one can only determine part of information on them. One can use a more accurate apparatus or instrument, or another method or more rigid theory to find the existence of the former, then search their variation law, and remove their influence. For the latter, one can only express them in the final expression of the results (see the following sections on the uncertainty).

1.5 Uncertainty of Measured Results

From the analysis above, it is not possible to view the measurement result absolute correct due to the existence of the random and systematic errors. There must be one or some indeterminate factors which affect the result. How to reasonably, adequately and scientifically evaluate and characterize the experiment result is the subject paid attention to by the international academic community. In 1992, the ISO (International Standardization Organization) issued the instruction to express the measurement uncertainty, since then the international guide rule for expression and evaluation of uncertainty is gradually established and accepted by most countries in the world. In order to help the experimenter form some basic concepts about uncertainty, we will introduce its definition, classification, property and method of expression.

1. Definition of uncertainty

Uncertainty, for short, U is the evaluation or measure of the range within which the true value (or the mean value) of the measured quantities is located. U determines the range of measured quantities near the mean value, namely $\bar{A} \pm U$, with a certain probability. Obviously the less the value of U, the higher the degree of the confidence or probability level is.

2. The classification and treatment of uncertainty

The uncertainty arises from different sources and forms various components, which are mainly divided to the followings.

(1) The uncertainty of kind A, evaluating the uncertainty by statistic method, symbolized by S_i (i denotes the i-th component). S_i mainly represents the uncertainty with random errors. For instance, in most measurements, we only take the standard deviation of mean value of sub-series of the measurement as the dominant term, that is

$$S_i = S_{\bar{A}} \qquad (1.5.1)$$

(2) The uncertainty of kind B, evaluating the uncertainty by non-statistic method, symbolized by u_j (j denotes the j-th component). We mainly consider the indeterminate systematic error as such a kind of uncertainty. For example, the instrument error Δ_I is directly related to such a kind of uncertainty. As we know that it gives the limit of error, in which the confidence probability is 100%. In order to match the standard of 68.3%, the Δ_I must be reduced by a factor C:

$$u_j = \frac{\Delta_I}{C} \qquad (1.5.2)$$

If the instrument errors follow the law of normal distribution (such as the physical balance) we take $C = 3$, while for uniform distribution (say the ruler) we take $C = \sqrt{3}$.

The total uncertainty U is the integration (synthesis) of all components of two kinds of uncertainties:

$$U = \sqrt{\sum_{i=1}^{m} S_i^2 + \sum_{j=1}^{n} u_j^2} \qquad (1.5.3)$$

where m, n are the number of components of kind A, B of uncertainties, respectively.

1.6 Expression of Direct Measured Results

For a direct measurement of a quantity A, if all other kinds of errors are eliminated or corrected the measured result should be expressed in the following form:

Part 1　Introduction to Laboratory Physics—About Measurement Errors, Data Treatment and Analysis

$$\begin{cases} A = \bar{A} \pm U \\ E = \dfrac{U}{\bar{A}} \times 100\% \\ \left(\begin{matrix}\text{confidence}\\ \text{probability}\end{matrix}\right) P = 68.3\% \end{cases} \qquad (1.5.4)$$

where U is the total integrated uncertainty; E is the fractional (relative or percentage) uncertainty. Note that the standard of confidence probability of $P = 68.3\%$ in which the measurement is taken must be attached in the final result. The meaning of the above form is that the true value of measured quantity A falls within the interval of $[\bar{A} - U, \bar{A} + U]$ with a confidence probability of 68.3%.

Example: Repetitively measured the length of an object using a venire caliper with 50 subdivisions (instrument error is $1/50 = 0.02$ mm), the measurement result of 6 times is listed as follows in unit of mm: 139.70, 139.72, 139.68, 139.70, 139.74, 139.72. Write down the expression of measurement result.

Answer: The instrument error $\Delta_1 = 0.02$ mm and its systemic errors follow the uniform distribution, so the uncertainty component of kind B can be obtained:

$$u_1 = \frac{\Delta_1}{C} = \frac{0.02}{\sqrt{3}} = 0.012 \text{ (mm)}$$

The mean value of L:

$$\bar{L} = \frac{1}{6}\sum_{i=1}^{6} L_i = \frac{1}{6}(139.70 + 139.72 + 139.68 + 139.70 + 139.74 + 139.72)$$
$$= 139.71 \text{ (mm)}$$

So the standard deviation of the measurement series:

$$S_{\bar{L}} = \sqrt{\frac{1}{6(6-1)}\sum_{i=1}^{6}(L_i - \bar{L})^2} = 0.009 \text{ (mm)}$$

The synthetic uncertainty:

$$U = \sqrt{S_{\bar{L}}^2 + u_1^2} = \sqrt{0.009^2 + 0.012^2} = 0.015 \text{ (mm)}$$

The final expression for the result:

$$\begin{cases} L = \bar{L} \pm U = (139.71 \pm 0.02) \text{ mm} \\ E = \dfrac{U}{\bar{L}} \times 100\% = \dfrac{0.02}{139.71} = 0.01\% \\ P = 68.3\% \end{cases}$$

1.7　Expression of Indirect Measured Results

In many scientific research and engineering applications, one often meets problems that need indirect measurement. Therefore, it is important to study the law for error transfer and expression method.

1. The mean value of an indirect measurement

Most indirect measurement quantity Y can be expressed in a function with quantities of direct measurement, in which the variable $x_i (i = 1,2,3,\cdots,n)$ denotes the different direct measured quantities (may be in different units or dimensions):

$$Y = f(x_1, x_2, x_3, \cdots, x_n) \tag{1.7.1}$$

The mean value of the indirect measurement must be calculated by substituting every mean value of the direct measurement into the function:

$$\overline{Y} = f(\overline{x}_1, \overline{x}_2, \overline{x}_3, \cdots, \overline{x}_n) \tag{1.7.2}$$

Note that the getting an average of Y corresponding to all combinations of the repetitive direct measurement must not become the approach to the mean value of a indirect measurement because sometimes the number of combinations is too big to calculate.

2. The uncertainty of an indirect measurement

For simplicity, by assuming that every direct measurement x_i is independent of each other, we can obtain the corresponding synthetic uncertainty U_i. It is proved that the uncertainty of an indirect measurement may be integrated from contribution of uncertainty of every direct measurement in the following form:

$$\begin{aligned} U &= \sqrt{\sum_{i=1}^{n} \left(\frac{\partial Y}{\partial x_i}\right)^2 U_i^2} \\ &= \sqrt{\left(\frac{\partial Y}{\partial x_1}\right)^2 U_1^2 + \left(\frac{\partial Y}{\partial x_2}\right)^2 U_2^2 + \cdots + \left(\frac{\partial Y}{\partial x_n}\right)^2 U_n^2} \\ &= \sqrt{\left(\frac{\partial f}{\partial x_1}\right)^2 U_1^2 + \left(\frac{\partial f}{\partial x_2}\right)^2 U_2^2 + \cdots + \left(\frac{\partial f}{\partial x_n}\right)^2 U_n^2} \end{aligned} \tag{1.7.3}$$

Here one may find that the total uncertainty could not be obtained by simply adding up every uncertainties of direct measurement. This is not only because sometimes x_i is not in the same unit or dimension. The uncertainty contribution of every direct measurement to the total uncertainty needs different weight $\partial f/\partial x_i$ — the partial derivative of the function with respect to x_i (to get a derivative when the other variables are considered as constants).

Sometimes it is convenient to calculate the relative uncertainty of indirect measurement first by

$$\begin{aligned} E &= \frac{U}{\overline{Y}} = \sqrt{\sum_{i=1}^{n} \left(\frac{\partial \ln Y}{\partial x_i}\right)^2 U_i^2} \\ &= \sqrt{\left(\frac{\partial \ln Y}{\partial x_1}\right)^2 U_1^2 + \left(\frac{\partial \ln Y}{\partial x_2}\right)^2 U_2^2 + \cdots + \left(\frac{\partial \ln Y}{\partial x_n}\right)^2 U_n^2} \\ &= \sqrt{\left(\frac{\partial \ln f}{\partial x_1}\right)^2 U_1^2 + \left(\frac{\partial \ln f}{\partial x_2}\right)^2 U_2^2 + \cdots + \left(\frac{\partial \ln f}{\partial x_n}\right)^2 U_n^2} \end{aligned} \tag{1.7.4}$$

which depends on the form of the function, for example, if the indirect measurement quantity

Part 1　Introduction to Laboratory Physics—About Measurement Errors, Data Treatment and Analysis

and direct measurement quantities is in the power, multiplication/division or square relation.

3. The final result expression of an indirect measurement

The final expression of an indirect measurement result is like that of a direct measurement one:

$$\begin{cases} Y = \bar{Y} \pm U \\ E = \dfrac{U}{\bar{Y}} \times 100\% \\ \left(\begin{matrix}\text{confidence}\\\text{probability}\end{matrix}\right) P = 68.3\% \end{cases} \quad (1.7.5)$$

Example 1: An indirectly measured quantity Y is linked by the two directly measured quantities X_1 and X_2 via $Y = f(X_1, X_2) = 2X_1 - X_2$, where X_1 and X_2 are independent of each other and the corresponding uncertainties U_1 and U_2 are known. Calculate the uncertainty of Y.

Answer:

$$U = \sqrt{\sum_{i=1}^{n=2} \left(\frac{\partial Y}{\partial X_i}\right)^2 U_i^2}$$

$$= \sqrt{\left(\frac{\partial f}{\partial X_1}\right)^2 U_1^2 + \left(\frac{\partial f}{\partial X_2}\right)^2 U_2^2}$$

$$= \sqrt{2^2 U_1^2 + U_2^2}$$

Example 2: In the simple sigle pendulum, we used the formula of $g = 4\pi^2 l/T^2$ to measure the acceleration of local gravity, in which the length of the pendulum l and the period T are measured as $\bar{l} = 70.6$ cm, $U_l = 0.2$ cm, $\bar{T} = 1.688$ s, $U_T = 0.007$ s. Calculate the final result.

Answer: The mean value of the acceleration of local gravity is

$$\bar{g} = \frac{4\pi^2 \bar{l}}{\bar{T}^2} = \frac{4 \times 3.141\,6^2 \times 0.706}{1.688^2} = 9.782 \ (\text{m/s}^2)$$

We first calculate the relative uncertainty of g:

$$E = \frac{U}{\bar{Y}} = \sqrt{\sum_{i=1}^{n} \left(\frac{\partial \ln g}{\partial x_i}\right)^2 U_i^2}$$

$$= \sqrt{\left(\frac{\partial \ln g}{\partial l}\right)^2 U_l^2 + \left(\frac{\partial \ln g}{\partial T}\right)^2 U_T^2}$$

$$= \sqrt{\left(\frac{1}{\bar{l}}\right)^2 U_l^2 + \left(\frac{2}{\bar{T}}\right)^2 U_T^2} = \sqrt{(E_l)^2 + (2E_T)^2}$$

$$= \sqrt{\left(\frac{0.2}{70.6}\right)^2 + \left(\frac{2 \times 0.007}{1.688}\right)^2} = 0.008\,8$$

We get the absolute uncertainty:

$$U = \bar{g} \cdot E = 9.782 \times 0.008\,8 = 0.086 \ (\text{m/s}^2)$$

So the final expression of the result is written as

$$\begin{cases} g = \bar{g} \pm U = (9.78 \pm 0.09) \text{ m/s}^2 \\ E = \dfrac{U}{\bar{g}} = 100\% = 0.88\% \\ \left(\begin{matrix}\text{confidence}\\\text{probability}\end{matrix}\right) P = 68.3\% \end{cases}$$

1.8 Significant Figures (Digits) and Their Algorithm

1. Least count and significant figures

In general, there are exact numbers and measured numbers (or quantities). Factors such as the 2 in $2\pi r$ are exact numbers. Measured numbers, as the name implies, are those obtained from measurement instruments and generally involve some errors or uncertainty.

2. The concept of significant figures

In experiment, a measured value or calculated result is often expressed by a set of figures or digits, namely the significant figures (significant digits). Every datum is composed of two parts: the credible figures and the doubtful figures, which is usually with a single digit and located in the end. Sometimes the doubtful figures are gotten by estimating, so it is natural that a measured datum possesses error to some extent. The degree of uncertainty of a number read from a measurement instrument depends on the quality of the instrument and the fineness of its measuring scale. From different instruments, the same physical quantity may be read in significant figures with different length of digits. For instance, a length of about 10 mm of an object may be measured to be 10.2 mm by a ruler, 10.18 mm by a venire caliper while 10.176 mm by a micrometer. The number of the significant figures and the location of the doubtful figure in a given unit depend on the least count of the instrument scale, which is the smallest subdivision on the measurement scale. This is the unit of the smallest reading that can be made without estimating. For example, the least count of a meter stick (ruler) is usually the millimeter (mm). We commonly say "the meter stick is calibrated in centimeters (numbered major divisions) with a millimeter least count".

In reporting experimentally measured values, it is important to read instruments correctly. When reading the value from a calibrated scale, only a certain number of figures or digits can properly be obtained or read. That is, only a certain number of figures are significant.

The significant figures of a measured value include all the numbers that can be read directly from the instrument scale, plus one doubtful or estimated number — the fractional part of the least count smallest division. Obviously, the length 10.2 mm read from a ruler (as measured from the zero end) is known to be between 10 mm and 11 mm. The estimated fraction is taken to be 2/10 of the least count (mm), so the doubtful figure is 2, giving 10.2 mm with three significant figures.

Note that, the greater the number of significant figures, the greater the reliability of the measurement the number represents. As mentioned above, the length may be obtained as 10.18 mm (four significant figures) by a venire caliper; it is also read as 10.176 mm (five significant figures) on anther instrument (micrometer) scale. The latter reading is from an instrument with a finer scale and gives more information and reliability.

In expressing a datum, zeros and the decimal point must be properly dealt with in determining the number of significant figures. For example, how many significant figures does 0.005 s have? What about 500.0 s? In such cases, the following rules are generally used to determine significance in treating and expressing the significant figures.

(1) In reading the figures, we must give the figures at least to the smallest subdivision of the instrument. The last figure may be zero, for example, a datum with four figures, 323.0 mm, in which zero occupies the place of the least accuracy digit, is expressed more accurately than 323 mm although in math 323.0 is as large as 323 numerically. Zeros at the beginning of a datum are not significant. They merely locate the decimal point. We start to count the significant figures from the first non-zero digit. For example, 0.054 3 m only has three significant figures (5, 4, and 3).

(2) Zeros within or at the end of a number are significant. For example, 400.4 s has four significant figures (4, 0, 0, and 4) while 4 000.0 s has five significant figures (4, 0, 0, 0, and 0).

(3) The change in location of decimal point and in unit can not change the number of significant figures. For example, 1.234 cm and 12.34 mm depict the same the significant figure in the different units.

(4) Some confusion may arise with whole numbers that have one or more zeros at the end without a decimal point. Consider, for example, 300 kg, where the zeros (called trailing zeros) may or may not be significant. In such cases, it is not clear which zeros serve only to locate the decimal point and which are actually part of the measurement (and hence significant). That is, if the first zero from the left (3<u>0</u>0 kg) is the estimated digit in the measurement, then only two digits are reliably known, and there are only two significant figures. Similarly, if the last zero is the estimated digit (30<u>0</u> kg), then there are three significant figures. This ambiguity can be removed by using <u>scientific (powers of 10) notation</u>: 3.0×10^2 kg has two significant figures. 3.00×10^2 kg has three significant figures. This procedure is also helpful in expressing the significant figures in large numbers. For example, suppose that the average distance from Earth to the Sun, 93 000 000 mi (1 mi = 1 609 m), is known to only four significant figures. This is easily expressed in powers of 10 notation: 9.300×10^7 mi.

3. Computation of measured quantities

Calculations are often performed with measured values, and error and uncertainty are "conveyed" or "transferred" by the mathematical operations such as, multiplication or division. That is, errors are carried through to the results by the mathematical operations. The error can be better expressed by statistical methods. However, a widely used procedure for estimating the uncertainty of a mathematical result involves the use of significant figures.

Before giving the computation rule for significant figures, it is worth noticing the famous "buckets effect" (the "cask effect" or the "effect of shortest slab") — the ability of a cask or barrel to fill the liquid such as wine depends on the shortest one from the slabs or boards that is used to truss the cask. In the computation, a reliable digit meets another reliable digit the computation result must be reliable. If the unreliable meets the reliable digit (or unreliable digit) the result becomes unreliable certainly. This generates a conclusion that the reliability of computation result of significant figures is determined by the datum with worst significant figures. One can see this in the following computation examples.

The number of significant figures in a measured value gives an indication of the uncertainty or reliability of a measurement. Hence, you might expect that the result of a mathematical operation can be no more reliable than the quantity with the least reliability, or smallest number of significant figures, used in the calculation. That is, reliability cannot be gained through a mathematical operation.

It is important to report the results of mathematical operations with the proper number of significant figures. This is accomplished by using rules for multiplication and division, addition and subtraction. To obtain the proper number of significant figures, one rounds the results off. The general rules used for mathematical operations and rounding as follows.

(1) Significant figures in calculations.

①When multiplying and dividing quantities, leave as many significant figures in the answer as there are in the quantity with the least number of significant figures.

②When adding or subtracting quantities, leave the same number of decimal places (rounded) in the answer as there are in the quantity with the least number of decimal places.

(2) Rules for rounding.

①If the first digit to be dropped is less than 5, leave the preceding digit as is.

②If the first digit to be dropped is 5 or greater, increase the preceding digit by one.

Notice that in this method, five digits (0, 1, 2, 3 and 4) are rounded down and five digits (5, 6, 7, 8 and 9) are rounded up. What the rules for determining significant figures mean is that the result of a calculation can be no more accurate than the least accurate quantity used. That is, you cannot gain accuracy in performing mathematical operations.

Part 1 Introduction to Laboratory Physics—About Measurement Errors, Data Treatment and Analysis

Example 1: 97.4 + 6.238 =? (Here the digit highlighted means unbelievable one.)

Answer:

$$\begin{array}{r} 97.\underline{4} \\ +\ 6.2\underline{38} \\ \hline 103.\underline{638} \end{array}$$

The final result is 103.6 (with 4 significant figures).

Example 2: 217 − 14.8 =?

Answer:

$$\begin{array}{r} 21\underline{7} \\ -\ 14.\underline{8} \\ \hline 20\underline{2.2} \end{array}$$

The final result is 202 (with 3 significant figures).

Example 3: 13.6 × 1.6 =?

Answer:

$$\begin{array}{r} 13.\underline{6} \\ \times\ \ 1.\underline{6} \\ \hline 8\underline{1}\ \underline{6} \\ 13\underline{6} \\ \hline 2\underline{1.7\ 6} \end{array}$$

The final result is 22 (with 2 significant figures).

Example 4: 2.453 × 6.2 =?

Answer:

$$\begin{array}{r} 2.45\underline{3} \\ 6.\underline{2} \\ \hline 490\underline{6} \\ 1471\underline{8} \\ \hline 15.\underline{2086} \end{array}$$

The final result is 15.2 (with 3 significant figures).

Example 5: 48 216 ÷ 123 =?

Answer:

$$\begin{array}{r} 392 \\ 12\underline{3}\overline{)48216} \\ 369 \\ \hline 1131 \\ 110\underline{7} \\ \hline 24\underline{6} \\ 24\underline{6} \\ \hline 0 \end{array}$$

The final result is 39\underline{2} (with 3 significant figures).

1.9 Treatment of Experimental Data — Arrangement, Demonstration and Analysis

In experiments, one often tries to study the variation laws of several physical quantities to reveal the inherent relationship among them, so it is of great importance to correctly record the data, reasonably classify the data and to get key parameters from plotting the relation between two quantities.

1. Tabulation method of data

Tabulation or listing is a necessary way for orderly recording the original data. It helps not only roughly to find variation law among the quantities, but also to check the rationality in the measurement data and in the calculation result. In tabulation, the following rules should be obeyed:

(1) At the top of a table the title or heading should be written;

(2) In every column the quantity name and the unit should be given in the firse place;

(3) In the table the original data are listed, sometimes some intermediate results are also inserted;

(4) If there exists a functional relation among the data, it is better to arrange the arguments or independent variables in ascending order or descending order.

Table 1.9.1 gives the voltage − current (Volt − Ampere) characteristic of a resistance R.

Table 1.9.1   The V − A characteristic of R

Voltage/V	1.00	2.00	3.00	4.00	5.00	6.00	7.00	8.00	9.00
Current/mA	0.50	1.02	1.49	2.05	2.51	2.98	3.52	4.00	4.48

2. Graphical representation of data

It is often convenient to represent experimental data in graphical form not only to report but also to obtain information. A graph is more visual and more intuitional than a table to reflect the relation between two quantities. The advantages of the graphic method are:

(1) The quantity relation and the variation law is represented by the graph;

(2) From the curve, one can evaluate or estimate a quantity that was not measured in the real experiment, and even extrapolate the quantity that is not plotted in the graph by the curve trend;

(3) From the slope, intercept even some characteristic points (say inflexion) of a graph one can obtain parameters that need to be found.

Quantities are commonly plotted using rectangular Cartesian axes (X and Y). The horizontal axis (X) is called the abscissa, and the vertical axis (Y) is called the ordinate. The location of a point on the graph is defined by its coordinates x and y, written (x, y), referenced to the origin O, the intersection of the X and Y axis. When plotting data, choose axis scales that are easy to plot and read. Also, the dots or data points should not be connected. Choose scales so that most of the graph paper is used. Scale units should always be included. It is also acceptable, and saves time, to use standard unit abbreviations, such as N for newton and m for meter. This will be done on subsequent graphs. With the data points plotted, draw a smooth line described by the data points. Smooth means that the line does not have to pass exactly through each point but connects the general areas of significance of the data points. The graph with an approximately equal number of points on each side of the line gives a "line of best fit". In cases where several determinations of each experimental quantity are made, the average value is plotted and the mean deviation or the standard deviation may be plotted as error bars. A smooth line is drawn so as to pass within the error bars. (Your instructor may want to explain the use of a French curve at this point.) Graphs should have the following elements:

(1) Each axis labeled with the quantity plotted;
(2) The units of the quantities plotted;
(3) The title of the graph on the graph paper (commonly listed as the y-coordinate versus the x-coordinate);
(4) Your name and the date;
(5) The title of the graph.

Fig. 1.9.1 gives an example of graph: the V − A characteristics (in fact a straight line), in which the abscissa denotes the voltage while the ordinate — the current. In real case, between the two set of data which one is chosen as the abscissa and the ordinate, depends on your convenience.

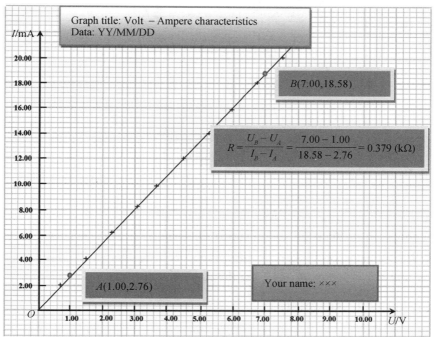

Fig. 1.9.1   The V − A characteristics curve plotted on the sheet of graph paper

After pointing the all quantities, one should search a straight line to "best fit" the data, in fact, not all data points are located on this line.

3. Method of successive difference

In the experiment, one may meet the linear relation between two variables x and y: $y = kx + b$, where k and b are called the slope and intercept of the straight line represented by this relation. Assuming x varies with almost the same interval and its errors are ignorable, one can use the <u>method of successive difference</u>. We regroup the variable y into two sets (the number of measurements should be an even number $2n$, where n is an integer):

$$y_1, y_2, y_3, \cdots, y_n; y_{n+1}, y_{n+2}, \cdots, y_{2n} \tag{1.9.1}$$

Find the difference between the corresponding terms:

$$\Delta y_1 = y_{n+1} - y_1, \Delta y_2 = y_{n+2} - y_2, \cdots, \Delta y_n = y_{2n} - y_n \tag{1.9.2}$$

And find the mean differendce:

$$\overline{\Delta y} = \frac{1}{n}\sum_{i=1}^{n}\Delta y_i = \frac{\Delta y_1 + \Delta y_2 + \cdots + \Delta y_n}{n}$$
$$= \frac{(y_{n+1} - y_1) + (y_{n+2} - y_2) + \cdots + (y_{2n} - y_n)}{n} \tag{1.9.3}$$

Thus the slope of the straight line can be obtained by

$$k = \frac{\overline{\Delta y}}{\Delta x} = \frac{1}{n}\frac{\Delta y_1 + \Delta y_2 + \cdots + \Delta y_n}{(x_{n+1} - x_1)}$$
$$= \frac{(y_{n+1} - y_1) + (y_{n+2} - y_2) + \cdots + (y_{2n} - y_n)}{n(x_{n+1} - x_1)} \tag{1.9.4}$$

The method has the following advantage over the averaging all the slopes with smallest interval made by dividing $\Delta y_1/\Delta x_1$, $\Delta y_2/\Delta x_2$, $\cdots$, $\Delta y_{2n-1}/\Delta x_{2n-1}$, where $\Delta y_i/\Delta x_i = (y_{i+1} - y_i)/(x_{i+1} - x_i)$ with $i = 1,2,3,\cdots,2n-1$:

$$k = \frac{1}{n}\sum_{i=1}^{2n-1}\frac{\Delta y_i}{\Delta x_i} = \frac{\Delta y_1 + \Delta y_2 + \cdots + \Delta y_n}{n\Delta x}$$
$$= \frac{(y_2 - y_1) + (y_3 - y_2) + \cdots + (y_{2n} - y_{2n-1})}{n\Delta x}$$
$$= \frac{(y_{2n} - y_1)}{n\Delta x} \tag{1.9.5}$$

In the procedure, almost all the quantities become useless except the first one and the last one. If the two data are with large errors, this may bring the inaccurate result although all $\Delta x_i = x_{i+1} - x_i (i = 1,2,3,\cdots,2n-1)$ are the same and accurate.

4. Method of least squares

Let $y = kx + b$ be the predicted equation of the best-fitting straight line for a set of data. The vertical deviation of the ith data point from this line is then $(y_i - y)$ (assuming that errors of x is ignorable). The principle of least squares may be stated as follows.

Part 1 Introduction to Laboratory Physics—About Measurement Errors, Data Treatment and Analysis

The "best-fitting" straight line is the one that minimizes the sum of the squares of the deviations of the measured y values from those of the predicted equation $y = kx + b$. That means

$$\sum_{i=1}^{n} (\Delta y_i)^2 = \sum_{i=1}^{n} (y_i - y)^2 = \sum_{i=1}^{n} [y_i - (kx_i + b)]^2 = \min \quad (1.9.6)$$

It suggests

$$\begin{cases} \dfrac{\partial}{\partial k} \sum\limits_{i=1}^{n} (\Delta y_i)^2 = -2 \sum\limits_{i=1}^{n} (y_i - b - kx_i) x_i = 0 \\ \dfrac{\partial}{\partial b} \sum\limits_{i=1}^{n} (\Delta y_i)^2 = -2 \sum\limits_{i=1}^{n} (y_i - b - kx_i) = 0 \end{cases} \quad (1.9.7)$$

If both equations are multiplied by $1/2n$ (n is the number of measurements), they become

$$\begin{cases} \dfrac{1}{n} \sum\limits_{i=1}^{n} (y_i - b - kx_i) x_i = 0 \\ \dfrac{1}{n} \sum\limits_{i=1}^{n} (y_i - b - kx_i) = 0 \end{cases} \quad (1.9.8)$$

We have

$$\begin{cases} \overline{xy} - b - k\overline{x_i^2} = 0 \\ \overline{y} - b - k\overline{x} = 0 \end{cases} \quad (1.9.9)$$

with

$$\overline{x} = \frac{1}{n} \sum_{i=1}^{n} x_i \quad \text{(mean value of } x\text{)}$$

$$\overline{x^2} = \frac{1}{n} \sum_{i=1}^{n} x_i^2 \quad \text{(mean value of squared } x\text{)}$$

$$\overline{y} = \frac{1}{n} \sum_{i=1}^{n} y_i \quad \text{(mean value of } y\text{)}$$

$$\overline{xy} = \frac{1}{n} \sum_{i=1}^{n} x_i y_i \quad \text{(mean value of product of } x \text{ and } y\text{)}$$

So we get the numerical values of the slope k and intercept b that minimize the sum of the squares of the deviations:

$$\begin{cases} k = \dfrac{\overline{x} \cdot \overline{y} - \overline{x \cdot y}}{\overline{x}^2 - \overline{x^2}} \\ b = \overline{y} - k\overline{x} \end{cases} \quad (1.9.10)$$

1.10 Exercises

* * *You must perform the following exercises before coming to the Lab!* * *

(1) There is a length measurement result which is expressed as
$$L = (25.78 \pm 0.05) \text{ mm } (P = 68.3\%)$$
Among the following descriptions, which one is correct?

a. The true value of the measured length is 25.73 mm or 25.83 mm.

b. The true value of the measured length is arranged from 25.73 mm to 25.83 mm.

c. Probability of the true value of the measured length located between 25.73 mm and 25.83 mm is 68.3%.

(2) Measure a resistance R with an identical accuracy and the data are listed as follows:
$$R(\Omega): 29.18, 29.14, 19.27, 19.25, 29.26$$
Find the mean value of R, standard deviation of measurement set and standard deviation of mean values of measurement set of R.

(3) Using a steel ruler repeatedly measure a length L with an identical accuracy:
$$L(\text{mm}): 10.2, 9.8, 10.5, 9.9, 10.1, 9.8, 10.2$$
The instrument error of the ruler is 0.5 mm and obeys the uniform distribution, please write out the last expression of the result.

(4) Using a timer (stopwatch) measure a time interval t to obtain the single measurement value $t = 20.20$ s. If the instrument error of the stopwatch is 0.1 s and assuming it obeys the normal distribution, please find the final expression of the result.

(5) The following functions give expressions for some indirect measurement ($P = 68.3\%$), please deduce their error propagation relations and find the final expression of the result.

a. $Y = A + 2B - 5C$, where $A = (38.206 \pm 0.001)$ cm, $B = (12.43 \pm 0.041)$ cm, $C = (8.340 \pm 0.008)$ cm.

b. $Y = \dfrac{4m}{\pi D^2 h}$, where $m = (236.124 \pm 0.002)$ g, $D = (2.346 \pm 0.005)$ cm, $h = (8.21 \pm 0.01)$ cm.

(6) Please revise the following expressions.

a. $N = (10.800 \pm 0.2)$ cm.

b. $Q = (1.612\ 48 \pm 0.287) \times 10^{-19}$ C.

c. $L = 12.0$ km ± 100 m.

d. $E = (1.93 \times 10^{11} \pm 6.76 \times 10^9)$ N/m^2.

(7) The dependence of resistance R on temperature T is as listed in the following data table (Table 1.10.1). Please draw the curve of R vs T on a sheet of graph paper and directly find the temperature coefficient of resistance using the slope.

Table 1.10.1 Dependence of R on T

$T/°C$	0.0	8.2	14.7	32.5	51.1	68.1	82.1	93.4
R/Ω	3.9	7.4	11.5	19.1	27.0	33.4	37.8	41.5

(8) The relation between coefficient of surface tension σ and temperature T (in Kelvin) is expressed in the following table (Table 1.10.2). Assuming that the function $\sigma = aT - b$ can represent the relation, please find a and b using:

①the method of successive difference;

②the least square method.

Compare the results with those directly determined from drawing on the sheet of graph paper.

Table 1.10.2 Dependence of T on σ

T/K	10.0	20.0	30.0	40.0	50.0	60
$\sigma/(N \cdot m^{-1})$	74.22	72.75	71.18	69.56	67.91	66.18

Part 2 Topics of Experiments

2.1 Measurement of Density

Common laboratory measurements involve the determination of the fundamental properties of mass and length. Most people are familiar with the use of scales and rulers or meter sticks. However, for more accurate and precise measurements, laboratory balances and vernier calipers or micrometer calipers are often used, particularly in measurements involving small objects.

In this initial experiment on measurement, you will learn how to use these instruments and what advantages they offer. Density, the ratio of mass to volume, will also be considered, and the densities of several materials will be determined experimentally.

1. Purpose

(1) To master some methods to measure the density of solid object;
(2) To make acquaintance with the usage of balance;
(3) To master the usage of vernier caliper and micrometer;
(4) To practice the treatment of significant figures and data analysis.

2. Principle

(1) For a regularly shaped object.

The density ρ of a substance is defined as the mass m per unit volume V (that is, $\rho = m/V$). Thus, the densities of substances or materials provide comparative measures of the amounts of matter in a particular (unit) space. Note that there are two variables in density — mass and volume. Hence, densities can be affected by the masses of atoms and/or by their compactness (volume). As can be seen from the defining equation ($\rho = m/V$), the SI units of density are kilogram per cubic meter (kg/m^3). However, measurements are commonly made in the smaller metric units of grams per cubic centimeter (g/cm^3), which can easily be converted to standard units. Density may be determined experimentally by measuring the mass and volume of a sample of a substance and calculating the ratio m/V. The volume of a regularly shaped object may be calculated from length measurements. For example, for cylinder, $V = A \times H$, where circular cross-sectional area $A = \pi r^2$, where r is the radius and H is the length of the cylinder. In fact, one can easier get the diameter D (and H) than r by a vernier caliper. So we have the expression for density of cylinder:

$$\rho = \frac{4m}{\pi D^2 H} \tag{2.1.1}$$

(2) For an irregularly shaped object.

Some objects float and others sink in a given fluid — a liquid or a gas. The fact that an object floats means it is "buoyed up" by a force equal to its weight. Archimedes, a Greek scientist, deduced that the upward buoyant force acting on a floating object is equal to the weight of the fluid it displaces. Thus, an object sinks if its weight exceeds that of the fluid it displaces. In this experiment, Archimedes' principle will be studied in an application: determining the densities or specific gravities of solid and liquid samples.

When placed in a fluid, an object either floats or sinks. This is most commonly observed in liquids, particularly water, in which "light" objects float and "heavy" objects sink. But the same effect occurs for gases. A falling object is sinking in the atmosphere, whereas other objects float.

Objects float because they are buoyant, or are buoyed up. That is, when submerged there must be an upward force that is greater than the downward force of the object's weight, and on release the object will be buoyed up and float. When floating, the upward buoyant force is equal to the object's weight. The upward force resulting from an object being wholly or partially immersed in a fluid is called the buoyant force. How the buoyant force arises can be understood by considering a buoyant object being held under the surface of a liquid. In this case, the buoyant force is balanced by the downward applied force and the weight of the block. It is not difficult to derive an expression for the magnitude of the buoyant force. This general result is known as Archimedes' principle: an object immersed wholly or partially in a fluid experiences a buoyant force equal in magnitude to the weight of the volume of fluid that it displaces. Thus the magnitude of the buoyant force depends only on the weight of the fluid displaced by the object, not on the weight of the object.

(3) Density of a heavy solid object ($\rho > \rho_0$).

When the object is wholly immersed in a given fluid — water, whose density ρ_0 is less than that of the object ρ, one can determine the buoyant force by measuring difference of the object's true weight in air w_0 and submerged weight in water w'_0 (Fig. 2.1.1), there is

$$F_b = w_0 - w'_0 = m_0 g - m'_0 g \tag{2.1.2}$$

according to Archimedes' principle, which is equal in magnitude to the weight of the volume of water that object displaces, thus

$$F_b = \rho_0 V g \tag{2.1.3}$$

where V is the object's volume, so we have

$$V = \frac{F_b}{\rho_0 g} = \frac{m_0 g - m'_0 g}{\rho_0 g} = \frac{m_0 - m'_0}{\rho_0} \tag{2.1.4}$$

Hence from measured object's true weight w_0 and its volume V we can determine its density ρ:

$$\rho = \frac{w_0}{gV} = \frac{m_0}{m_0 - m_0'}\rho_0 = \frac{w_0}{w_0 - w_0'}\rho_0 \qquad (2.1.5)$$

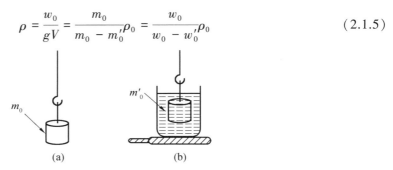

Fig. 2.1.1 Measurement of mass (a) in air and (b) when the object is wholly immersed in the water

Finding the density of water at a room temperature ρ_0 from the table in appendix and twice measuring the object's weights in air w_0 and in water w_0', one can get the density of an object that is larger than that of water.

(4) Density of a light solid object ($\rho < \rho_0$).

If the water's density ρ_0 is larger than that of the measured object ρ, which means the object can't sink or wholly be immersed in water on the natural state, one can first measure the weight of object in air w_0, fix a heavy object onto it with a filament (the mass of which can be ignored) and only immerse the heavy object into the water leaving the measured object in the air, then get the total weight w_0', later on measured the total weight when all objects immersed into the water w_0'', thus the buoyant force exerted by the measured object may be determined by measuring difference between w_0' and w_0'' (Fig. 2.1.2), there is

$$F_b = w_0' - w_0'' = (m_0' - m_0'')g = \rho_0 V g = \frac{\rho_0 w_0}{\rho} = \frac{\rho_0 m_0 g}{\rho} \qquad (2.1.6)$$

So we have

$$\rho = \frac{w_0}{w_0' - w_0''}\rho_0 = \frac{m_0}{m_0' - m_0''}\rho_0 \qquad (2.1.7)$$

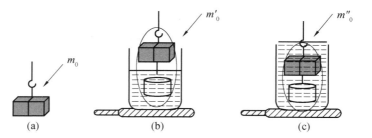

Fig. 2.1.2 Measurement of mass (a) in air, (b) only immersing the heavy object and leaving the measured object in air, and (c) both immersing the heavy object and the measured object

Hence, finding the density of water at room temperature ρ_0 from Table 2.1.1 in the appendix, individually measuring w_0, w_0' and w_0'', one may obtain the density of the object that lighter than the density of water.

(The students should deduce the expression of uncertainty for the two cases mentioned above).

Table 2.1.1 The dependence of water's density on temperature

Temperature $T/℃$	Density $\rho_0/(\mathrm{kg \cdot m^{-3}})$	Temperature $T/℃$	Density $\rho_0/(\mathrm{kg \cdot m^{-3}})$	Temperature $T/℃$	Density $\rho_0/(\mathrm{kg \cdot m^{-3}})$
0	999.87	12	999.52	24	997.32
1	999.93	13	999.40	25	997.07
2	999.97	14	999.27	26	996.81
3	999.99	15	999.13	27	996.54
4	1 000.00	16	998.97	28	996.26
5	999.99	17	998.80	29	995.97
6	999.97	18	998.62	30	995.67
7	999.93	19	998.43	31	995.37
8	999.88	20	998.23	32	995.05
9	999.81	21	998.02	33	994.72
10	999.73	22	997.80	34	994.40
11	999.63	23	997.57	35	994.06

3. Experiment contents

(1) Regularly shaped object.

①Determine the masses of measured objects using the physical scale and digital electronic scale, respectively, recording the instrument errors (please fill in the Table 2.1.2 in the report).

②Determine the size of the regularly shaped object (height H and diameter D) using vernier caliper and micrometer, recording their instrument errors (please fill in the Table 2.1.3 in the report).

③Calculate the density's uncertainty and give the final expression.

(2) Object whose density is larger than water's.

①Determine the buoyant force by measuring difference of the object's true weight in air m_0 and submerged weight in water m_0' (please fill in the Table 2.1.4 in the report).

②Calculate its density.

(3) Object whose density is less than water's.

①Measure the weight of object in air m_0 (please fill in the Table 2.1.5 in the report).

②Fix a heavy object onto it with a filament (ignore its influence on the measurement) and only immerse the heavy object into the water leaving the measured object in the air, then measure the total weight m_0', later on measure the total weight m_0'' when all objects immersed into the water (please fill in the Table 2.1.5 in the report).

③Calculate the density using the equation mentioned above.

4. Discussion

(1) How to measure density of a liquid with known liquid (say water)?

(2) How to eliminate the effect of unbalance on the measurement in a scale with two unbalanced arms?

(3) Please try to explain the effect when an air bubble fixed on the surface of an immersed object.

Part 2　Topics of Experiments

Experimental Report

Measurement of Density

Name:　　　　　　　Student ID No.:　　　　　Score:

Date:　　　　　　　　Partner:　　　　　　　　Teacher's signature:

1. Purpose

2. The experimental data and processing

(1) For brass cylinder (Table 2.1.2, Table 2.1.3).

Table 2.1.2　Measurement of mass

		m/g	
	m	m/g	
		m/g	
Instrument error $\Delta m/g$		Mean value	

Table 2.1.3　Measurement of height and diameter

Measurement times	D/cm		H/cm	
1				
2				
3				
4				
5				
Mean-value				
Deviation s	s_D/cm		s_H/cm	
Δ_I	Δ_D/cm		Δ_H/cm	
u	u_D/cm		u_H/cm	
U	U_D/cm		U_H/cm	

$\rho =$

$U_\rho =$

Final expression:

(2) For aluminum sample (Table 2.1.4).

Table 2.1.4 Measurement of masses in air and immersed in water

m_0 (in air)	m_0/g	m_0' (immersed in water)	m_0'/g
	m_0/g		m_0'/g
	m_0/g		m_0'/g

$\rho =$

(3) For wax sample (Table 2.1.5).

Table 2.1.5 Measurement of masses

	1	2	3	4	5
m_0/g					
m_0'/g					
m_0''/g					

$\rho =$

3. Discussion

2.2 Measurement of Young's Elastic Modulus

1. Purpose

(1) To study the measurement approach of Young's modulus by pulling the metal wire;
(2) To master the principle of "optical lever" to measure a very slight change in length;
(3) To master the method of successive difference to treat data.

2. Principle

In an object of solid state under the action of external force a change in shape may occur, it is called deformation. In a given limit, when the force ceases the deformation disappears. Such a deformation is termed as elastic deformation. While the force is too large, the deformation could not be eliminated — a residual deformation left, it is named as "plastic deformation".

The simplest deformation in a thin rod is the stretching out (elongation) and drawing back (contraction). The ratio of the net elongation of a length Δl to the original length l, $\Delta l/l$, is called the strain. In the elastic limit, according to the Hooker's law on the elasticity, when the external force increase to F' from original force F, the rod stretch Δl is in direct proportional to the increment of force, we have a strain relation for a rod:

$$\frac{F' - F}{S} = E \frac{\Delta l}{l} \tag{2.2.1}$$

where S denotes the area of cross section of the thin rod, the coefficient E is termed as Young's modulus or modulus of elasticity, which is one of the important material parameters. E is often used to describe an ability of a material to resist the deformation. For a thin circular rod, say the steel wire, the cross section $S = \pi d^2/4$, where d is the diameter of rod, so we have a relation:

$$E = \frac{4(F' - F)}{\pi d^2} \frac{l}{\Delta l} \tag{2.2.2}$$

which is used to measure the Young's modulus of steel.

It should be pointed out that usually the amount Δl is too small or subtle to perceive, so how to detect such trivial change becomes a key point of the experiment. Later we will introduce the principle of "optical lever" to assist the measurement.

Principle of "optical lever". The optical lever is such a mechanical structure that used to amplify the change in length. A subtle change Δl is reflected through change in position of light spot reflected by a mirror which is mounted on the supporting structure with a tail with tip of certain length, as shown in Fig. 2.2.1 and Fig. 2.2.2. With no change ($\Delta l = 0$), the reflected light spot need to be adjusted to just coincide with the starting point of the incident light, which is often called as the self-collimation state, which means that both the incident light and reflected light are perpendicular to the mirror (parallel with its normal line). In fact, we cannot guarantee such a state rigidly due to the natural misalignment of the positions of the telescope and the ruler in the experimental setup (please see the following descriptions).

When the mirror rotates by a small angle θ caused by the elongation of the rod, the mirror-linked position of reflected light spot will be change to another place. If the position change can be read by a ruler, the subtle change in l is reflected into the change in readings on it. We first derive the relation of amplification of the structure.

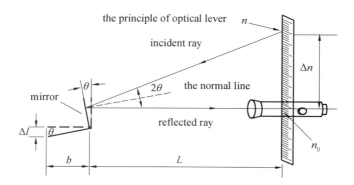

Fig. 2.2.1 The mechanical structure of an "optical lever" and its amplifying principle

When a small strain is generated by adding the kilogram weights to the steel wire, the change will be reflected into the change of reflecting angle via the descending of a holder fixed on the lower end of steel wire. Assuming that the setup initially is adjusted to self collimation state, the ray from the same position with the observer's eye (from a telescope) can be caught by the observer. When the normal line of the mirror is rotated by a small angle θ caused by the change in length of steel wire, we have

$$\tan\theta \approx \theta = \frac{\Delta l}{b} \quad (2.2.3)$$

where b is length of a tail of the optical lever and called as the length of one of the beams of "optical lever", which is defined as a distance from holder to mirror. The reflected ray pointing to the direction with 2θ with respect to the incident ray can be detected. From the Fig. 2.2.1 we find

$$\tan 2\theta \approx 2\theta = \frac{n - n_0}{L} \quad (2.2.4)$$

where $n - n_0$ denotes the change in readings on ruler through telescope, L is the distance from the mirror to the telescope. Through above equations, eliminating the angle, we have

$$n - n_0 = \frac{2L}{b}\Delta l \quad (2.2.5)$$

This is the famous formula for an "optical lever", by which a small quantity is amplified like the force's role is amplified by a lever. It is easily to know that the magnification is $2L/b$. As the length of another beam L must be larger than b, the magnification must be larger than unity. There is a factor "2" because the light undergoes the path back and forth that doubles the magnification. We should notice that the condition for validity of all the formulae is the angle θ must be small enough and it is better that initially we set $\theta_0 = 0$. In fact, it is difficult to let the initial angle θ_0 be zero but one should guarantee it small enough. It can be proved that the amplification formula is still valid with small nonzero θ_0. At last, we obtain the expression for measuring the Young's modulus:

$$E = \frac{8lL}{\pi bd^2} \frac{F' - F}{n - n_0} \qquad (2.2.6)$$

where l denotes the original length of the steel wire between upper and lower holders. Hence we can get E by measuring the beam length b (with mirror), the diameter of steel wire d, the original length of the steel wire and the distance L from the mirror to ruler near the telescope, and by reading out the increment of reading change on the ruler $n - n_0$ with change in the increment of value of the load (the kilogram weights) $F' - F$.

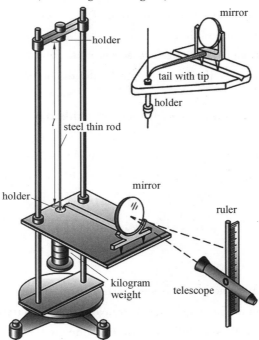

Fig. 2.2.2 The experiment setup for Young's modulus. In the inset the real structure of an optical lever and the way for mounting on the holder is given in detail

3. Experiment contents

(1) Adjust the experimental setup (Fig. 2.2.2) to the state, on which the measurement can be performed.

It should be pointed out that the ultimate aim is that one should observe the reading change on a ruler via a telescope so the order of adjustment is very important. The ruler (must be on the correct state) must clearly be reflected by the mirror (on the correct state) via a telescope (on the correct state). Any disorder in adjustment would influence the measurement quality. To fast make all parts in order one must roughly let the every part on the correct state. The experimenter should pay more attention to the adjustment process.

(2) Measure the change caused by the change in strain, and various geometric parameters.

Measure the reading change on ruler observing through telescope in the order of adding and subtracting load by 1 kg and fill in the blanks as given in Table 2.2.1 in experimental report. Notice that during the measurement, the mirror, telescope and ruler are not allowed to touch. Any shake may cause the sudden transition of the data, so adding and subtracting the weights should be performed carefully.

Using different instrument, repetitively measure the original length l of the steel wire (why not the whole length?), the beam length b, the diameter of steel wire d, the distance L (Table 2.2.2 in report).

(3) Treat the data using the method of successive difference (Table 2.2.3 in report) and calculate Young's modulus E.

(4) Calculate the uncertainty (ignoring the weight errors) and correctly express the result using the following expression (please self check it and give the deduction procedure):

$$\frac{U_E}{\overline{E}} = \sqrt{\frac{U_L^2}{\overline{L}^2} + \frac{U_l^2}{\overline{l}^2} + \frac{U_b^2}{\overline{b}^2} + \frac{4U_L^2}{\overline{d}^2} + \frac{U_{n-n_0}^2}{(\overline{n-n_0})^2}} \qquad (2.2.7)$$

4. Discussion

(1) The measurement principle is based on the ideal assumption or approximation that the rotation angle $\theta - \theta_0$ must be small enough to avoid the measurement errors. What factors will increase $\theta - \theta_0$? How does the initial angle θ_0 influence the measurement? Please try to analyze it and give some methods to control the effects.

(2) Illustrate the Hooker's law briefly.

Part 2 Topics of Experiments

Experimental Report

Measurement of Young's Elastic Modulus

Name: Student ID No.:   Score:

Date: Partner: Teacher's signature:

1. Purpose

2. The experimental data

Table 2.2.1 Measurement of reading change with adding and subtracting the weights

Net weight /kg	Readings on ruler via telescope/cm		
	Adding	Subtracting	Average
1			
2			
3			
4			
5			
6			

Table 2.2.2 Recording the parameters

L/cm	$\bar{L}$/cm	l/cm	$\bar{l}$/cm	b/cm	$\bar{b}$/cm	d/mm	$\bar{d}$/mm

Table 2.2.3 The successive method to treat data

$(F' - F)$/N	$(n - n_0)$/m	$\overline{(n - n_0)}$/m
$F_4 - F_1 = 3 \times 9.8$	$n_4 - n_1 =$	
$F_5 - F_2 = 3 \times 9.8$	$n_5 - n_2 =$	
$F_6 - F_3 = 3 \times 9.8$	$n_6 - n_3 =$	

3. Data processing (calculating the modulus)

$$E = \frac{8lL}{\pi b d^2}\left(\frac{F' - F}{n - n_0}\right) =$$

$$\frac{U_E}{\overline{E}} = \sqrt{\frac{U_L^2}{\overline{L}^2} + \frac{U_l^2}{\overline{l}^2} + \frac{U_b^2}{\overline{b}^2} + \frac{4U_L^2}{\overline{d}^2} + \frac{U_{n-n_0}^2}{(\overline{n - n_0})^2}} =$$

$U_E =$

Final expression:

4. Discussion

2.3 Measurement of Liquid's Viscosity

1. Purpose

(1) To master some knowledge about the viscosity of liquid;

(2) To study the falling-sphere method to determine the viscosity coefficient of a liquid;

(3) To master the usage of reading/travelling microscope.

2. Principle

(1) Viscous fluid flow.

Any liquid possesses viscosity to a certain degree. It is considerably easier to pour water out of a container than to pour honey. This is because honey has a higher viscosity than water. In a general sense, viscosity refers to the internal friction of a fluid. It's very difficult for layers of a highly viscous fluid to slide past one another. Likewise, it's difficult for one solid surface to slide past another if there is a highly viscous fluid, such as soft tar, between them.

When an ideal (non-viscous) fluid flows through a pipe, the fluid layers slide past one another with no resistance. If the pipe has a uniform cross section, each layer has the same velocity, as shown in Fig. 2.3.1(a). In contrast, the layers of a viscous fluid have different velocities, as Fig. 2.3.1(b) indicates. The fluid has the greatest velocity at the center of the pipe, whereas the layer next to the wall doesn't move because of adhesive forces between molecules and the wall surface.

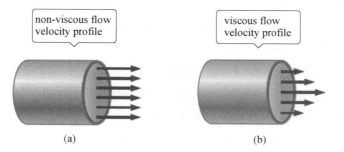

Fig. 2.3.1 (a) The particles in an ideal (non-viscous) fluid all move through the pipe with the same velocity;

(b) in a viscous fluid, the velocity of the fluid particles is zero at the surface of the pipe and increases to a maximum value at the center of the pipe

To better understand the concept of viscosity, consider a layer of liquid between two solid surfaces, as shown in Fig. 2.3.2. The lower surface is fixed in position, and the top surface moves to the right with a velocity v under the action of an external force F. Because of this motion, a portion of the liquid is distorted from its original shape, $ABCD$, at one instant to the shape $AEFD$ a moment later. The force required to move the upper plate and distort the liquid is proportional to both the area A in contact with the fluid and the speed v of the fluid. Further, the force is inversely proportional to the distance d between the two plates. We can express these proportionalities as $F \propto Av/d$. The force required to move the upper plate at a fixed speed v is therefore

$$F = \eta \frac{Av}{d} \qquad (2.3.1)$$

where η (the lowercase Greek letter eta) is the <u>coefficient of viscosity</u> of the fluid. The SI unit of viscosity is $N \cdot s/m^2$ or $Pa \cdot s$.

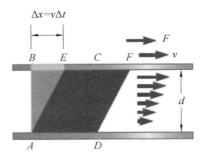

Fig. 2.3.2   Characteristics of viscous force

The units of viscosity in many references are often expressed in $dyne \cdot s/cm^2$, called 1 poise, in honor of the French scientist J. L. Poiseuille. The relationship between the SI unit of viscosity and the poise is

$$1 \quad poise = 0.1 \quad N \cdot s/m^2 = 0.1 \quad Pa \cdot s$$

Small viscosities are often expressed in centipoise (cp), where $1 \text{ cp} = 10^{-2}$ poise.

(2) Method of falling sphere (ball-dropping method).

As a small sphere (ball) falls through an unbounded viscous liquid, there are three forces acting on it: force of gravity mg, buoyant force f_b, and viscous force f_v. The last two are the resistive forces. If the ball is small and moving in an infinite and isotropic liquid with an adequately small velocity, the viscous force can be expressed by the famous Stokes's law:

$$f_v = 6\pi \eta v r \qquad (2.3.2)$$

where v, r are the velocity, the radius of the ball, respectively. It is clear that the gravity force is downward while the buoyancy force and the viscous force are upward. At the instant the sphere begins to fall, the force of friction is zero because the speed of the sphere is zero. As the sphere accelerates, its speed increases and so does viscous force f_v. Finally, at a speed called <u>the terminal speed</u> v_t, the net force goes to zero. This occurs when the net upward force balances the downward force of gravity $mg = f_b + f_v$, namely

$$\frac{4}{3}\pi r^3 \rho g = \frac{4}{3}\pi r^3 \rho_0 g + 6\pi \eta v_t r \tag{2.3.3}$$

where ρ and ρ_0 are the density of the sphere and the liquid, respectively. Thus we have

$$\eta = \frac{2}{9}\frac{(\rho - \rho_0)g r^2}{v_t} \tag{2.3.4}$$

That means one can easily determine the viscosity coefficient of the liquid by measuring the sphere size and its terminating velocity in falling through the liquid. This is the main physical idea of the sphere falling method.

In real case, the experiment is carried out usually in a limited container such as the measuring tank (metering cylinder), the wall's effect of the container must be considered. Let the diameter of the metering cylinder to be D and the height of the liquid to be H. For the case when the ball is falling in the center of the cylinder in vertical position, the modified formula can be used:

$$\eta = \frac{1}{18}\frac{(\rho - \rho_0)g d^2}{v_t\left(1 + 2.4\frac{d}{D}\right)\left(1 + 3.3\frac{d}{2H}\right)} \tag{2.3.5}$$

where d is the diameter of the sphere. Due to the fact that $d/H \ll 1$, one can ignore the corresponding modified term.

Usually we mark two lines (Fig. 2.3.3): the upper one (in some place below the liquid surface to guarantee the falling velocity reaches the terminating velocity v_t) and the lower one (high above the bottom of the cylinder to eliminate the bottom's effect), between which the velocity is viewed as the uniform (identical). If the distance L between the two lines and the time t the sphere undergoes are measured, the velocity is determined by $v_t = L/t$. Finally, we end up with

$$\eta = \frac{1}{18}\frac{(\rho - \rho_0)g t d^2}{L\left(1 + 2.4\frac{d}{D}\right)} \tag{2.3.6}$$

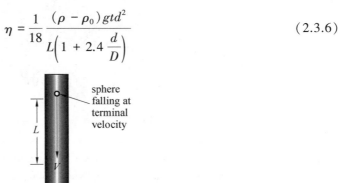

Fig. 2.3.3 The limited container: metering cylinder

(3) Reading microscope.

In the experiment, reading microscope is useful to measuring the sphere's diameter d, as shown in Fig. 2.3.4.

The main adjustments and measurement procedures are listed below:

①Adjust the ocular to clearly catch the crosshairs inside;

②Put the object on the measurement table and lower the microscope tube (just let the objective closely above the sphere but not allowed to touch it);

③Raise the microscope tube in order clearly to see the sphere in visual field of the microscope, in order to reduce parallax error, it is necessary to make sure that the image of the sphere should lie at the plane of the crosshairs;

④If necessary, adjust the trend of the crosshair to be sure that the horizontal line of the crosshair is parallel to the translation direction (Fig. 2.3.5);

⑤It is notable that in order to avoid hysteresis error, we should rotate the micrometer drum in the same direction during the measurement.

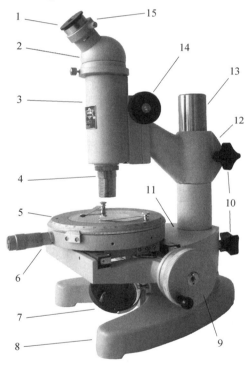

Fig. 2.3.4 A reading microscope

1—the ocular; 2—prism part; 3—the tube of microscope; 4—the objective; 5—measuring rotational table with holder of sample; 6—screw micrometer; 7—the reflecting mirror; 8—the instrument support; 9—the transverse micrometer drum; 10—the fixing knobs; 11—the measurement platform; 12—the holder of microscope; 13—the main support column; 14—the adjusting screw for raising and lowering telescope tube; 15—the circular ocular holder

Part 2   Topics of Experiments

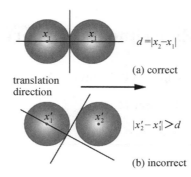

Fig. 2.3.5   The relations between translation direction and trend of the crosshairs in the measurement

3. Experimental Contents

(1) Measure the diameters ($d = |x_1 - x_2|$) of five steel balls by using the reading microscope. Each ball is measured in three different directions.

(2) Measure the time t that the ball undergoes through the two marks of the cylinder:

①Release the ball into the central hole of the foil on the top of cylinder;

②When the bottom of the ball reaches the mark at the top of the cylinder, start the stopwatch;

③When the bottom of the ball reaches the mark at the bottom of the cylinder, stop the stopwatch.

4. Discussion

(1) When a small sphere follows on the heel of another sphere falling along the same direction (more accurately the same track), it trends to chase the former, please try doing the experiment and explain the reason. May the sphere behind catch up with the foregoing one? Is there any change occurring in the falling process? How to measure the falling time? How to determine the viscosity coefficient?

(2) When the two identical spheres abreast fall in a viscous liquid, will the falling time be larger than the falling time when one sphere falling individually. Someone says that "it will be larger, because the cross section doubled, the resistance will double, too", do you agree with this viewpoint?

(3) Explain why the big migratory birds (for example the group or flock of swans and wide geese) fly in a row or lambdoid shape. Why the leader-bird must be strong enough to conduct the birds? Why the weak birds often crowded in the middle to avoid from lagging and being breaking ranks?

Experimental Report

Measurement of Liquid's Viscosity

Name:　　　　　　　　　Student ID No.:　　　　　　　Score:

Date:　　　　　　　　　　Partner:　　　　　　　　　　Teacher's signature:

1. Purpose

2. The experimental data and processing

(1) Fulfill the data table below (Table 2.3.1).

Table 2.3.1　Measurement of the diameters and times

No. of balls	x_1/mm	x_2/mm	d/mm	$\bar{d}$/mm	t/s	T/℃	η/(Pa·s)
1							
2							
3							
4							
5							

The diameter of steel ball is defined as $d = |x_1 - x_2|$. $\bar{d}$ is the average of d; t is the falling time between the two marks of the cylinder; T is the temperature of the liquid; η is the viscosity coefficient and can be calculated according to the expression:

$$\eta = \frac{1}{18} \frac{(\rho - \rho_0) g t \overline{d}^2}{L\left(1 + 2.4\dfrac{\overline{d}}{D}\right)}$$

(Gravitational acceleration g = 9.806 6 m/s^2; the steel sphere's density ρ = 7.700 × 10^3 kg/m^3; the oil's density ρ_0 = 0.965 0 × 10^3 kg/m^3; the distance between the two marks L = 0.200 0 m; the diameter of cylinder D = 0.08 m.)

(2) Plot the curve of viscosity coefficient η versus temperature T on graph paper.

3. Discussion

2.4 Measurement of the Rotational Inertia of a Rigid Body

1. Purpose

To master the method of determining the rotational inertia of a rigid body using the torsional pendulum.

2. Principle

One of the instruments for measuring the rotational inertia is a torsional pendulum, whose schematic structure is shown in Fig. 2.4.1. On the vertical axis of the pendulum, a ribbon-shaped helical spring is fixed to generate a restoring torque. On the top of the axis, a variety of measurable rigid body can be mounted. To minimize the friction, the bearings are used between the axis and the holder. A bubble level is also fixed on the holder to guarantee the perpendicularity of the axis to the water level (horizontal plane).

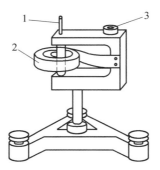

Fig. 2.4.1 Schematic structure of torsional pendulum
1—the axis; 2—the ribbon-shaped spring; 3—the bubble level

When we rotate the measured body an angle φ from the balance position, due to the deformation of spring a moment of restoring force M is generated, which forces the body to move toward the equilibrium position. Without any external moment of force, the body will make a periodic motion about the equilibrium position. Within the elastic limit of the spring M is directly proportional to the angle φ. Thus

$$M = -K\varphi \qquad (2.4.1)$$

where the proportional coefficient K is called as the torsional coefficient. The motion is a swing on the fixed axis, where there is only one degree of freedom — the angular displacement φ. The angular acceleration is proportional to the moment M, so we have

$$M = I\beta = -K\varphi \qquad (2.4.2)$$

where I denotes the rotational inertia of the rigid body. Thus

$$\beta = \ddot{\varphi} = -\frac{K}{M}\varphi = -\omega^2\varphi \qquad (2.4.3)$$

which means the motion of the torsional pendulum is a kind of angular harmonic motion, whose solution is in the form of

$$\varphi = A\cos(\omega t + \theta_0) \tag{2.4.4}$$

where A is the angular amplitude of the rotational harmonic motion; θ_0 is the initial phase; ω is the angular frequency of the motion. The oscillatory period of the swing can be expressed as

$$T = \frac{2\pi}{\omega} = 2\pi\sqrt{\frac{I}{K}} \tag{2.4.5}$$

We have

$$I = \frac{KT^2}{4\pi^2} \tag{2.4.6}$$

In order to get the rotational inertia of a rigid body I, we should measure the oscillation period T and the torsional coefficient K. In our experiment, K can be obtained first by measuring the swing period T_0 of a regular rigid body (its calculated rotational inertia I_0), i.e., there is

$$K = I_0 \frac{4\pi^2}{T_0^2} \tag{2.4.7}$$

In the lab, some rigid bodies are provided such as the hollow metal cylinder, a solid plastic cylinder, wood ball, and the thin metal rod with two metal blocks freely movable on it to verify the parallel axis theorem.

The setup for measuring the rotational inertia of a rigid body may be divided into two parts: the torsional pendulum and the measuring instrument for detecting/controlling/displaying data. The opto-electric detector consists of infrared emitter and receiver, which converts the optical signal into electric one. The signal is sent into a counter on the control panel while the timer on it controls the starting and ending time.

3. Experiment contents

(1) Be familiar with the structure of torsional pendulum, the usage method, and the control panel. Adjust the foot screw of the pendulum to let the bubble located in the center of the level.

(2) Measure and record the masses and geometrical parameters of all rigid bodies, including the two blocks fixed on the thin metal rod.

(3) Mount the metal loading disc on to the axis and measure the average swing period $\overline{T}_0$. Onto the loading disc coaxially place the solid plastic cylinder, whose rotational inertia I'_1 can be calculated theoretically, and measure the average swing period $\overline{T}_1$. By using I'_1, T_0 and T_1, the torsional coefficient K can be calculated from (please derive it by yourself)

$$K = 4\pi^2 \frac{I'_1}{\overline{T}_1^2 - \overline{T}_0^2} \quad (\text{N} \cdot \text{m}) \tag{2.4.8}$$

(4) Respectively measure the rotational inertias of the solid plastic cylinder, the hollow

metal cylinder, wood ball and the thin metal rod and compare them with theoretical computed results, calculate the percentage difference.

(5) Symmetrically fix the two movable blocks in the thin rod in the distance from the rotation axis 5.00 cm, 10.00 cm, 15.00 cm, 20.00 cm and 25.00 cm, measure the respective periods T_i to verify the parallel axis theorem for rotational inertia. Plot the curve of rotational inertia I vs d^2 on graph paper.

(Note: Please fulfill the Table 2.4.1 – 2.4.3 in the experimental report.)

4. Discussion

(1) In the experiment, the center of a rigid body may not coincide with the axis of rotation. What it influences on the measurement result if it happens? Does the result become larger than that in standard case? What about the effect if the torsional pendulum is not well adjust at the water level?

(2) In the real case because of the fiction's existence, the amplitude of oscillation decays. Is there any changes in the period? What effect it may exert on the measurement?

Experimental Report

Measurement of the Rotational Inertia of a Rigid Body

Name: Student ID No.:   Score:

Date: Partner: Teacher's signature:

1. Purpose

2. Data recording and processing

(1) Fulfill Table 2.4.1.

Table 2.4.1 Record the periods of measured rigid bodies

		Metal loading disc	Solid plastic cylinder	Hollow metal cylinder	Wood ball	Thin metal rod
Period/s	1					
	2					
	3					
	4					
	5					
Average period/s		$\bar{T}_0$	$\bar{T}_1$	$\bar{T}_2$	$\bar{T}_3$	$\bar{T}_4$

($\bar{T}_0$ denotes the average period for the loading disc; while $\bar{T}_1$ and $\bar{T}_2$ for the solid plastic cylinder and hollow metal cylinder on the loading disc, respectively; $\bar{T}_3$ and $\bar{T}_4$ for wood ball and thin metal rod, respectively.)

Calculate the theoretical result of the rotational inertia of solid plastic cylinder $I'_1 = \frac{1}{8}m\bar{D}_1^2 =$ _____ (kg · m²).

Calculate the torsional coefficient $K = 4\pi^2 \dfrac{I'_1}{\bar{T}_1^2 - \bar{T}_0^2} =$ _____ (N · m).

(2) Fulfill Table 2.4.2.

Table 2.4.2 Measured rotational inertia

Object	Mass /kg	Geometric dimensions /m	Average Period /s	Theoretic rotational inertia /(kg·m^{-2})	Measured rotational inertia /(kg·m^{-2})	Percentage difference
Metal loading disc	—	—	$\bar{T}_0$	—	$I_0 = \dfrac{I'_1 \bar{T}_0^2}{\bar{T}_1^2 - \bar{T}_0^2}$ =	—
Solid plastic cylinder		$\bar{D}_1$	$\bar{T}_1$	$I'_1 = \dfrac{1}{8} m \bar{D}_1^2$ =	$I_1 = \dfrac{K \bar{T}_1^2}{4\pi^2} - I_0$ =	
Hollow metal cylinder		$\bar{D}_{ex}$ $\bar{D}_{in}$	$\bar{T}_2$	$I'_2 = \dfrac{1}{8} m (\bar{D}_{ex}^2 + \bar{D}_{in}^2)$ =	$I_2 = \dfrac{K \bar{T}_2^2}{4\pi^2} - I_0$ =	
Wood ball		$\bar{D}_3$	$\bar{T}_3$	$I'_3 = \dfrac{1}{10} m \bar{D}_3^2$ =	$I_3 = \dfrac{K}{4\pi^2} \bar{T}_3^2$ =	
Thin metal rod		l	$\bar{T}_4$	$I'_4 = \dfrac{1}{12} m l^2$ =	$I_4 = \dfrac{K}{4\pi^2} \bar{T}_4^2$ =	

Note: The diameter of the solid plastic cylinder is $\bar{D}_1$; the inside and outside diameters of hollow metal cylinder are $\bar{D}_{in}$ and $\bar{D}_{ex}$, respectively; the diameter of the wood ball is $\bar{D}_3$; the length of the thin metal rod is l. The mass and geometric size of any object are given in the appendix.

(3) Fulfill Table 2.4.3.

Table 2.4.3. To verify the parallel axis theorem for rotational inertia

$d/(\times 10^{-2}\text{m})$		5	10	15	20	25
Period T_i/s	1					
	2					
	3					
	4					
	5					
	Averaged T					
$I/(\times 10^{-4}\text{kg}\cdot\text{m}^{-2})$						
$d^2/(\times 10^{-4}\text{m}^2)$		25	100	225	400	625

Note: In the table, I is the rotational inertia of two movable blocks with $I = \dfrac{KT^2}{4\pi^2} - I_4$ where I_4 is the rotational inertia of the thin metal rod.

(4) Based on the data of Table 2.4.3, plot the curve of rotational inertia I versus d^2 on graph paper.

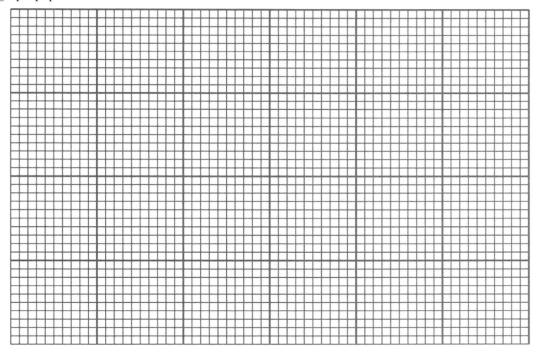

3. Discussion

4. Appendix: The masses and geometric quantities of rigid bodies

Solid plastic cylinder: diameter $\bar{D}_1 = 100.00$ mm, mass $m_1 = 712.9$ g.

Hollow metal cylinder: outside diameter $\bar{D}_{ex} = 100.00$ mm, inside diameter $\bar{D}_{in} = 94.00$ mm, mass $m_2 = 704.9$ g.

Wood sphere: diameter $\bar{D}_3 = 132.00$ mm, mass m_3 marked on it.

Thin metal rod: length $l = 610.00$ mm, mass $m_4 = 133.9$ g.

Movable block: mass $M = 239.0$ g, length $l = 33.00$ mm, diameter $d = 35.00$ mm.

2.5 Collision-Shooting Experiment

As common phenomena in nature, collisions between objects, from subatomic scales to astronomic scales, are of the most important research topics in physics.

In this experiment, the study on the collision between two metal spheres, as well as the related simple pendulum motion before collision and the projectile motion after collision, has been applied to solve the corresponding practical problems, further learn the laws of mechanics, understand mechanical principle, and improve the problem-solving ability.

1. Purpose

(1) To study the phenomenon and law of the collision;
(2) To solve practical problems using the law of kinematics and the law of conservation of mechanical energy;
(3) To evaluate the energy loss in the process of collision.

2. Principle

(1) Pendulum motion.

In the gravitational field, the potential energy of an pendulum ball with mass m has an increment of

$$\Delta E_p = mgh_0 \tag{2.5.1}$$

when it is lifted up by vertical distance h_0.

(2) The law of conservation of mechanical energy.

For isolated conservative system, no work done by external force or non-conservative force, the total mechanical energy consisting of potential energy and kinetic energy remains constant. When the pendulum ball reaches lowest level, it has a kinetic energy, which is equal to gravitational potential energy, given by

$$\frac{1}{2}mv^2 = mgh_0 \tag{2.5.2}$$

(3) Collision.

Refers to an isolated event in which two or more bodies exert strong forces on each other for a relative short time. The "head-on collision" means that the velocities of two colliding objects have directions parallel/antiparallel to the connection of their centers of mass. Other than that, collisions are called "oblique collision".

(4) The conservation of momentum.

The total momentum of the two objects remains unchanged during the collision. When two objects with identical mass undergo "head-on elastic collision" (elastic collision refers to the collision without mechanical energy loss in the process of collision), the kinetic energy of pendulum ball will be totally transferred into the stationary target ball:

$$\frac{1}{2}mv_0^2 = \frac{1}{2}mv^2 \tag{2.5.3}$$

(5) Horizontal projectile motion.

The motion equation for an object moving horizontally with initial velocity v_0, regardless of the air resistance, is given by

$$\begin{cases} x_0 = v_0 t \\ y = \dfrac{1}{2} g t^2 \end{cases} \quad (2.5.4)$$

where t is the traveling time since the object is launched; x_0 is the horizontal distance; y is the vertical distance; g is gravitational constant. Eliminate t in Eq.(2.5.4), we have

$$v_0^2 = \dfrac{g}{2y} x_0^2 \quad (2.5.5)$$

Considering Eq.(2.5.1) ~ (2.5.5), the relationship between x_0 and h_0 is given by

$$h_0 = \dfrac{x_0^2}{4y} \quad (2.5.6)$$

In this experiment, x_0 is the theoretical horizontal distance, x is the actual horizontal distance by measurement, the energy loss due to the collision is described as follows:

$$\begin{cases} E_{\text{before}} = \dfrac{1}{2} m v^2 = m g h_0 \\ E_{\text{after}} = \dfrac{1}{2} m v_0^2 = \dfrac{mg}{4y} \bar{x}^2 \\ \eta = \dfrac{\Delta E}{E} = \dfrac{E_{\text{before}} - E_{\text{after}}}{E_{\text{before}}} = \dfrac{4 y h_0 - \bar{x}^2}{4 y h_0} = 1 - \left(\dfrac{\bar{x}}{x_0} \right)^2 \end{cases} \quad (2.5.7)$$

3. Instruments

The experimental setup is shown in Fig. 2.5.1.

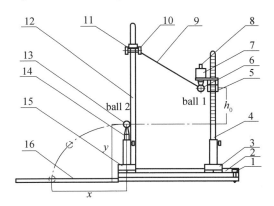

Fig. 2.5.1 The setup for collision experiment

1—adjusting screw; 2—guide way; 3—slide block; 4—column support; 5—scale plate; 6—pendulum ball; 7—electromagnet; 8—armature screw; 9—string; 10—locating screw; 11—adjusting knob; 12—column support; 13—target ball; 14—support; 15—slide block; 16—target surface

4. Experiment contents

(1) Adjust the level of guide way, keep it horizontal by adjusting screw.

(2) The mass of target ball is considered as the same mass of pendulum ball. The diameter of target ball is considered as the same diameter of pendulum ball, and the value of the diameter is taken as a constant, $D = 2$ cm.

(3) Have the pendulum ball attached to the electromagnet by magnetic attraction, adjust the height of pendulum ball to h_0, while keeping the string straight.

(4) Adjust the length of string, so that the pendulum ball could head-on collide with the target ball when the pendulum ball swings to the lowest level. Measure the height of the target ball y.

(5) Run the collisions experiment for three times, measure the corresponding horizontal traveling distance of target ball x, calculate the average distance $\bar{x}$, compared with x_0 (Eq.(2.5.6)), complete Table 2.5.1 in the experimental report.

(6) Calculate the energy loss due to the collision, complete Table 2.5.2 in the report.

5. Discussion

(1) If two balls with different masses have the same momentum, do they also have the same kinetic energy? If they are different, which one has the greater kinetic energy?

(2) If the target ball is not placed, can the pendulum ball reach its original height when it swings back? Explain the reason.

Experimental Report

Collision-Shooting Experiment

Name:　　　　　　　　Student ID No.:　　　　　　　Score:

Date:　　　　　　　　 Partner:　　　　　　　　　　Teacher's signature:

1. Purpose

2. The experimental data and process

(1) Measure the corresponding horizontal traveling distance (Table 2.5.1, Table 2.5.2).

Measure the corresponding horizontal traveling distance of target ball x, find the average distance $\bar{x}$, compared with x_0 (Eq.(2.5.6)); measure $y = \underline{\hspace{2cm}}$ cm; plot two curves in one graph ($\bar{x}$ versus h_0 and x_0 versus h_0)

Table 2.5.1　Measure the corresponding horizontal traveling distance

h_0/cm	x/cm			$\bar{x}$/cm	$x_0 = \sqrt{4h_0 y}$
	x_1	x_2	x_3		(Eq.(2.5.6))
13.00					
12.00					
11.00					
10.00					
9.00					
8.00					
7.00					
6.00					
5.00					
4.00					

Calculate the energy loss due to the collision and plot the experimental data in a graph (η versus h_0).

Table 2.5.2 Calculate the energy loss

h_0/cm	13.00	12.00	11.00	10.00	9.00	8.00	7.00	6.00	5.00	4.00
x/cm										
x_0/cm										
$\eta = \Delta E/E$ (Eq.(2.5.7))										

(2) Plot two curves in this graph ($\bar{x}$ versus h_0 and x_0 versus h_0).

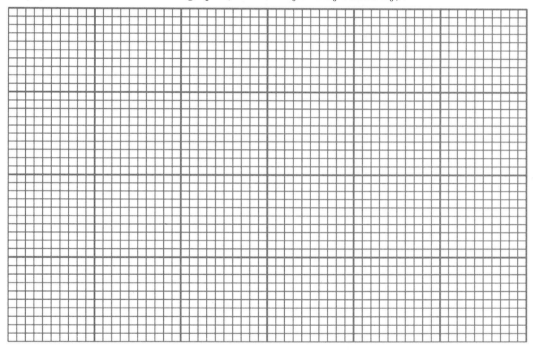

(3) Plot the experimental data in this graph (η versus h_0).

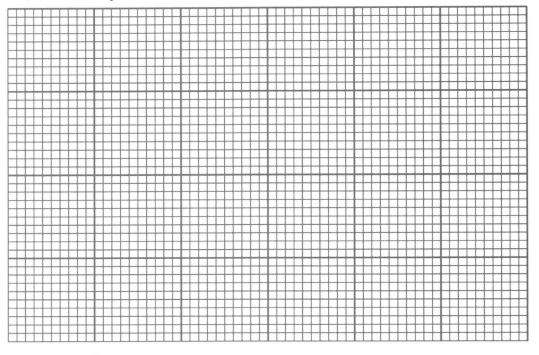

3. Discussion

2.6 Standing Waves on a String

A wave is the propagation of a disturbance or energy. When a stretched string or cord is disturbed, the wave travels along the string with a certain speed. Upon reaching a fixed end of the string, the wave is reflected back along the string. For a continuous disturbance, the forward waves interfere with the backward reflected waves, and a standing wave pattern is formed under certain conditions. These standing-wave patterns can be visually observed. The visual observation and measurement of standing waves serve to provide a better understanding of wave properties and characteristics.

1. Purpose

(1) To produce a standing wave and measure the wavelength;
(2) To analyze the corresponding experimental data;
(3) To prove the laws of vibration in a stretched string.

2. Principle

Waves in a stretched string are transverse waves, that is, the displacement of a particle on the spring is perpendicular to the direction of propagation. While in longitudinal waves, for example, in sound waves, the displacement of the particle is in the direction of propagation. When a stretched string is disturbed, the wave travels along the string with a speed V (m/s) that depends on the tension T inside the string and its line density μ with the relation:

$$V = \sqrt{\frac{T}{\mu}} \qquad (2.6.1)$$

A traveling wave also can be characterized by its wavelength λ (m), the frequency f(Hz), and the wave speed V (m/s) with

$$V = \lambda f \qquad (2.6.2)$$

By substituting Eq.(2.6.2) into Eq.(2.6.1), the relation among wavelength λ, tension T and line density μ can be obtained:

$$\lambda = \frac{1}{f}\sqrt{\frac{T}{\mu}} \qquad (2.6.3)$$

In order to prove Eq.(2.6.3), we usually take the common logarithm of both sides:

$$\lg \lambda = \frac{1}{2}\lg T - \frac{1}{2}\lg \mu - \lg f \qquad (2.6.4)$$

In the Eq.(2.6.4), if f and μ are fixed to constant, the $\lg \lambda$ and $\lg T$ will have the linear relation. Changing the value of tension T and measuring the corresponding wavelength λ, we can plot a curve for $\lg \lambda$ vs $\lg T$, in fact the curve should be a straight line with a slope of 1/2, indicating that $\lambda \propto T^{1/2}$, that is, the wavelength varies directly as the square root of the tension. On the other hand, if line density μ and tension T are kept constant, we can alter the frequency f and measure λ, then plot curve for the $\lg \lambda$ vs $\lg f$, the resulting curve should be a straight line with a slope of -1, that means $\lambda \propto f^{-1}$.

How to produce a standing wave and measure the wavelength is the important part in this experiment. For a continuous disturbance, the forward traveling waves interfere with the oppositely moving reflected waves, and a standing wave pattern is formed under certain conditions. These standing-wave patterns can be visually observed as shown in Fig. 2.6.1. Note that part of the "particles" on the string are stationary, the corresponding positions of which are called <u>nodes</u>. While those points for the particles with maximum displacements are called <u>antinodes</u>. The length of one loop of the standing wave is equal to one-half the wavelength of standing wave. <u>Stable nodes indicate the formation of a standing wave!</u>

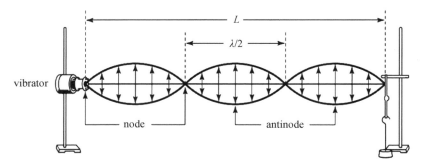

Fig. 2.6.1 The standing wave with nodes and antinodes

Since the string is fixed at each end, a standing wave must have a node at each end. As a result, only an integral number of half wavelengths may "fit" into a length L of the string, there is

$$L = n\left(\frac{\lambda_n}{2}\right) \quad \text{or} \quad \lambda_n = \frac{2L}{n}, \quad n = 1,2,3,\cdots \quad (2.6.5)$$

In order to calculate the wavelength, we should: ① Measure the distance L between two fixed ends; ② Count loops (n is the number of loops); ③ Calculate wavelength with Eq.(2.6.5). Obviously, the length of one "loop" of the standing wave is equal to half wavelength.

So the visual observation of standing waves provides a better way to measure the wavelength. In this experiment, the electrically driven string vibrator can operate in a certain frequency ranging from 0 Hz to 200 Hz. By changing the length L, the driving frequency and the tension of the string, one can get standing waves with different patterns.

3. Instruments

The vibrator, copper string, weights and hanger and a balance are important things in the experiment. The structure of experimental setup for studying standing wave is shown in Fig. 2.6.2.

Part 2　Topics of Experiments

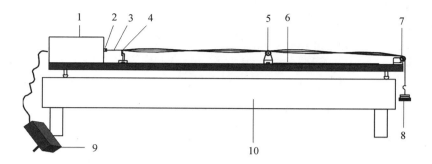

Fig. 2.6.2　Setup for standing waves

1—vibration generator; 2—vibrator; 3—string; 4—movable pointer; 5—bridge (ponticello); 6—ruler; 7—pulley; 8—weights and hanger; 9—electric plug; 10—table

A copper string is attached to the vibrator, and running over a movable pointer, bridge and fixed pulley. A traveling wave can be produced by the vibrator and reflected by the fixed end (bridge). The standing wave may be found between the vibrator and the bridge. The frequency can be read from the vibration generator, and the tension can be altered by adding and subtracting weights on the weight hanger.

4. Experiment contents

(1) Produce a standing wave (practice).

①Fix the tension and the frequency. Add several weights (total six weights) to the weight hanger. Turn on the vibrator and tune the frequency to a certain value like 120 Hz.

②To practice producing standing waves with different patterns by changing the length of string. Slide the bridge gently from the pulley to the vibrator, you will see some standing-waves with difference patterns (different number of loops).

③When a standing wave is not perfect, slide the bridge slightly back and forth until all the antinodes are with maximum amplitude and all nodes are obviously stationary.

(2) Measure the wavelength λ with different tensions T and complete Table 2.6.1 in the experimental report.

①Measure one weight on a balance, and record it as m_1. Put this weight on the hanger. Choose driven frequency $f = 120$ Hz. The value of the mass of the weight hanger has been tagged on the vibration generator, find it and record it as m_0.

②To produce a standing wave with two loops (between the nearest node from the vibrator and the bridge).

③Move the movable pointer to the nearest node and measure the distance between the bridge and the movable pointer. Record the distance as L.

④Repeat above procedures with more weights (adding weights one by one). Fill in Table 2.6.1 in the report.

59

(3) Measure the wavelength with given frequency (six weights suspended), and complete Table 2.6.2 in the report.

①Suspend six weights on the string.

②Produce a standing wave (with two loops) with a given frequency. Measure the distance between the bridge and the movable pointer (the nearest node from the vibrator). Record the distance as L. Complete Table 2.6.2 in the report.

5. Discussion

(1) Do the ends of a fixed string form knots or antinodes?

(2) What is the relationship between the linear density and the wave speed of the transverse wave of the string?

Experimental Report

Standing Waves on a String

Name: Student ID No.:   Score:

Date: Partner: Teacher's signature:

1. Purpose

2. The experimental data and processing

(1) Measure the wavelength with varying tension and complete Table 2.6.1.

Driven frequency $f =$ _____ Hz.

The value of the mass of the weight hanger $m_0 =$ _____ g.

Table 2.6.1 Measure the wavelength with varying tension

Number of weights	1	2	3	4	5	6
Suspended mass of weights $m/$g						
Total mass $(m + m_0)/$g						
The length for two loops $L^*/$cm						
Number of loops measured n						
Wavelength $\lambda/$cm						
Tension (in gram)** T						
lg λ						
lg T						

* When you do this step, actually, you produce a standing wave with three loops, measure the distance from the bridge 5 to the node nearest the vibrator. The fixed-end nodal point at the vibrator is not used because in vibrating up and down, it is not a "true" node. That means you just measure the length for two loops.

** In fact, the tension equals to the gravity force from the weight, namely $T = m'g$, where m' is the total mass of weight; g is the local gravity acceleration. For convenience, T may be expressed in terms of mass in gram (g) or kilogram (kg), that is if $m' = m + m_0 = 60$ g $= 60 \times 10^{-3}$ kg, then $T = m'g = 60 \times 10^{-3}$ kg. Thus lg $T =$ lg $m' +$ lg $g =$ lg $60 +$ lg $10^{-3} +$ lg g. Actually as the constants the last two terms (the one for unit conversion of mass and the other one for gravity acceleration g) only make a contribution to the intercept of the straight line of lg λ vs lg T, not to the slope!

Plot the experimental data on a graph paper for lg λ versus lg T and calculate the slope of the line with graphic method (details see Section 1.9 in Part 1).

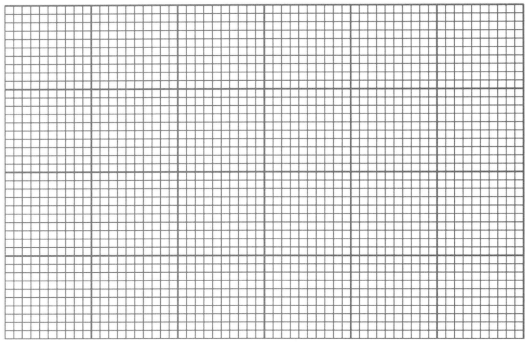

(2) Measure the wavelength with different frequency (suspended six weights) and complete Table 2.6.2.

The total mass of the weight hangers and hanger $m + m_0 = $ _____ g.

Table 2.6.2 Measure the wavelength with different frequency

Driven frequency f/Hz	75	80	85	90	95	100	105
The length for two loops L/cm							
Number of loops measured n							
Wavelength λ/cm							
lg λ							
lg f							

Plot the experimental data on a graph of lg λ versus lg f and calculate the slope of the line with graphic method (details see Section 1.9 in Part 1).

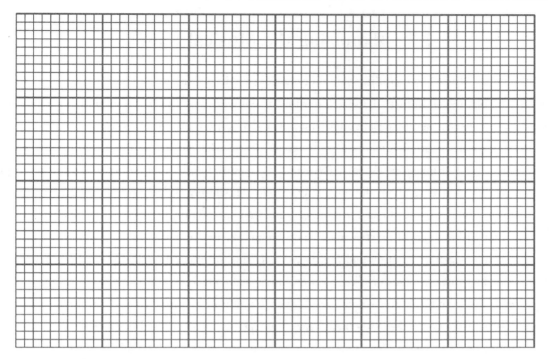

3. Discussion

2.7 Exploring the Musical Sounds

What is Music? As said in some dictionaries: music is "the art of combining sounds with a view to beauty of form and the expression of emotion and the art of organizing tones to produce a coherent sequence of sounds intended to elicit an aesthetic response in a listener". Both of these descriptions give us a pretty good idea of what music is, but neither really goes to the heart of the question. Music is, in fact, very difficult to completely and satisfactorily describe. When talking about the music, everyone can share a little understanding and feeling of his own. Music belongs to such a kind of thing that spreads its endless charm. Someone may accept that "like a sort of mysterious cosmic language, music may describe everything in the world, convey various emotion and feeling, it is easier for music to make a person infected by the joy or happiness, anger or rage, depression or sadness, relaxation or tension". Music lets you touch the grandness and solemnness, even tragedy of a thing, to feel the lingering and helplessness, to taste the sweetness and bitterness. Someone believes that music is a wondering "magic pen": it draws you a picture or image, sets you a scene, it unwinds a scroll of colorful painting. Music can show you the beauty and ugliness of a thing, the strength and softness. It may bathe you with the sunshine and immerse you into the darkness and threat. Music may depress you but it also can cheer you up by providing you endless strength.

Music, in fact, is a complex of the musical tones or good sounds, which spreads or is generated from variety of sources like musical instruments with different pitches, soundness and duration in a certain rhythm. Sometimes, it likes a stream flowing silently and peacefully. Sometimes, it becomes a big river that roars wildly running far away.

Why does a musical tone sound good? From where its charm comes? What is the nature of consonance or harmony in a combination of some musical tones in physics? This experiment will guide you to explore the mystery and secret of the musical sounds.

1. Purpose

(1) To master physical description method of a sound to study its characteristics in time and frequency domain and understand the concept of timbre;

(2) To study the frequency components of a musical sound and their relations and explore the pleasing reason to the ear;

(3) To explore the consonance of some combined musical sounds by considering their frequency ratios;

(4) To be familiar to functionality of the LabVIEW-based analyzing tool and digital storage oscilloscope.

2. Principle

(1) Vibration and wave propagation.

A sound is the mechanically vibrating wave, which may propagate in a certain medium like air, water and metal. In the medium, the sound may propagate in form of longitudinal or transverse wave. The audible range of sound frequencies spans from 20 Hz to 20 000 Hz. A sound may be complex in construction since it may be combined with some other sounds. The simplest sound wave is sine or cosine wave with a fixed frequency, also is viewed as a basic component, in other words by combination of components any wave can be synthesized.

A sine wave propagating along a certain direction (say x axis) with a fixed frequency can be described by a simple harmonic wave in time and space:

$$u(x,t) = A_0\cos(kx + \omega t + \varphi) \quad (2.7.1)$$

where A_0 denotes the vibration amplitude. k, ω and φ stand for the value of wave vector, frequency and the initial phase of vibration, respectively. In fact, k and ω determine the spatial and temporal periodicity via $k = 2\pi/\lambda$ and $\omega = 1/(2\pi T)$ with λ and T representing the wavelength and period of the vibrating wave. The two temporal and spatial quantities are linked through wave velocity, namely

$$v = f\lambda = 2\pi\omega\lambda \quad (2.7.2)$$

Fig. 2.7.1 - 2.7.2 gives schematically picture of a transverse simple harmonic wave distributed in space and the time.

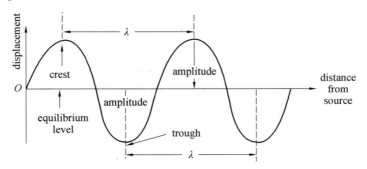

Fig. 2.7.1   Distribution of sine vibration in space at a fixed time

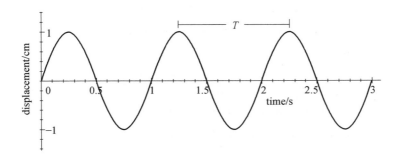

Fig. 2.7.2   Variation of sine vibration in time at a fixed point of observation

Usually, a musical sound or tone is a complex vibration with a certain (quasi-) periodicity, which corresponds to a central frequency (sometimes called as fundamental frequency). The concepts of pitch, loudness, duration and timbre are often used to describe a musical sound, which are directly linked the physical quantities such the central frequency, amplitude, time that the wave exists and the peculiarity or uniqueness of a sound reflected in temporal and frequency domain in fact. Actually, the sounds from different source like musical instruments or the vocal organs of living species sound differently, which corresponds to the different physics information and picture.

(2) Fourier components, standing waves and modes.

According to the theory of French scientist Jean Baptiste Joseph Fourier, any signal can be considered as a superposition of plenty of simple harmonic waves with various amplitude, different frequency and different phase. From this viewpoint, a musical sound may written as a vibration with form of combined simple harmonic waves:

$$u(x,t) = \sum_{n=1} A_{0n} \cos(k_n x + \omega_n t + \varphi_n), \quad n = 1, 2, 3, \cdots \quad (2.7.3)$$

where n is a integer. The nth wave corresponds to a frequency component. We often meet a sound from some musical instrument, in which $\omega_n = n\omega_0$, where ω_0, $\omega_n(n \neq 1)$ are termed as the fundamental frequency and the harmonic frequency, which actually forms a family of harmonics. It should be pointed out that the harmonics with $\omega_n(n \neq 1)$ has a synonym of "overtone" and the frequency of a strongest component is called as "main frequency". The fundamental frequency may coincide with main frequency sometimes.

The difference in the way of variation with time and the construction of frequency components decide the timbre of a musical sound.

The musical sound may be directly related to the standing wave and vibration modes.

On a vibrating string, it is easier to observe the propagation of forward and backward waves and the formation of the standing wave when the reflection caused by a bridge occurs. Standing wave has important characteristics that it demonstrates several special points in space, namely the nodes (where there are no vibration) and antinodes (where the amplitudes reach the maximum).

Fig. 2.7.3 shows some standing waves (modes) supported by a stretched segment of string with a certain tension. To satisfy the boundary condition, the length of the string L must be integer multiple of the half wavelength, namely $L = n\lambda/2$ (n is an integer), accordingly the frequency of each standing wave is

$$f_n = \frac{v}{\lambda_n} = n\frac{v}{2L}, \quad n = 1, 2, 3, \cdots \quad (2.7.4)$$

For the string with a tension σ and line density ρ, the velocity of a wave vibrated transversely can be expressed as $v = \sqrt{\sigma/\rho}$. Thus we have

$$f_n = \frac{v}{\lambda_n} = \frac{n}{2L}\sqrt{\frac{\sigma}{\rho}}, \quad n = 1, 2, 3, \cdots \quad (2.7.5)$$

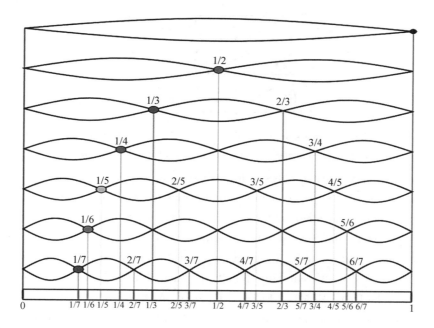

Fig. 2.7.3 The supported standing waves on a segment of string with a certain tension, both end of string is fixed

While for the case of longitudinal wave like the vibration of the air column in a wind instrument such as flute or trumpet, the standing wave are also formed inside. Sometimes we call this vibration as dilatational wave or the density varying wave. For the case of semi-enclosed cavity, in which one end is open another enclosed, the frequency corresponding to a supported mode is

$$f_n = \frac{v}{\lambda_n} = (2n-1)\frac{v}{4L}, \quad n = 1, 2, 3, \cdots \quad (2.7.6)$$

A fixed n means a certain mode. While for the case of open cavity, the frequency of supported modes is

$$f_n = \frac{v}{\lambda_n} = n\frac{v}{2L}, \quad n = 1, 2, 3, \cdots \quad (2.7.7)$$

Here the velocity of the wave is the sound speed in air, which is decided by

$$v = \sqrt{\frac{\gamma RT}{m}} = v_0\sqrt{1 + \frac{t}{273.15}} = 331.45\sqrt{1 + \frac{t}{273.15}} \quad (2.7.8)$$

where T and t denote the temperature in Kelvin and Celsius degree. γ, R, m stand for the air specific heat ratio, gas constant and average mass of air molecules, respectively. At the room temperature the sound speed is about 334 m/s. As an example, Fig. 2.7.4 gives the amplitude distribution of the first three modes ($n = 1, 2, 3$) supported by a semi-enclosed cavity.

From the expressions above one can see that for the open cavity and semi-enclosed cavity some frequency like fundamental frequency even the first overtone frequency is different.

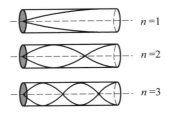

Fig. 2.7.4   The first three vibration modes ($n = 1, 2, 3$) supported by a semi-closed cavity, the schematic for the vibrational amplitude distribution

(3) The relationship between notes in a musical scale.

From a simple point of view, we can say that a musical scale is a special discrete set of notes that seem to belong together or be a linked sequence of tones. Internationally, if the pitch of key note is fixed, the corresponding frequencies of notes in this musical scale are also fixed. We will assume that we are dealing with a musical scale that has "C" as its key note or the scale is in the key of C. We can sing the all the musical notes in the diatonic scale in the key of C: C(Do), D(Re), E(Mi), F(Fa), G(Sol), A(La), B(Ti), C'(Do'). They are represented by the white keys on a piano keyboard, which may correspond to a definite frequency definition like $f_{Do} = 262$ Hz, $f_{Re} = 294$ Hz, $f_{Mi} = 330$ Hz, $f_{Fa} = 349$ Hz, $f_{Sol} = 392$ Hz, $f_{La} = 440$ Hz, $f_{Ti} = 494$ Hz, $f_{Do'} = 523$ Hz. Obviously, the interval between any pair of notes should obey a certain rule or regularity.

In fact, the West commonly thinks that an integer ratio exists among the notes of a musical scale, which was first discovered by the ancient Greek philosopher Pythagoras. He found that for a stretched string, when using a half of original length and keeping the same tension, the pitch would be enhanced twice. When plucking the this string and original one together it give a pleasing chord, in which the frequency ratio is 2/1, like playing Do and Do', in fact the interval between them is a perfect octave. He compared the tones from the section that was 2/3 of the length of the entire original string to one from the entire string. Again, he found that the two tones were pleasing when sounded together. We know that they are perfect fifth apart, which can be described by an integer ratio of 3/2 in frequency. In the same way, he found another pleasing combination of tone linked with intervals of perfect fourth and major third, which correspond to the integer frequency ratio of 4/3 and 5/4. Let's look more closely at the ratios he discovered, beginning with number 1, and writing them as 1/1, 5/4, 4/3, 3/2, 2/1. Their decimal equivalents are 1.000, 1.250, 1.333, 1.500, 2.000. They represent five important musical intervals: the perfect unison, major third, perfect fourth, perfect fifth, and octave. From them Pythagoras was able to devise a musical scale called the pentatonic scale.

The musical interval expresses the relationship between the pitches and hence frequencies between a pair of notes. The neighboring notes in a scale may be semitone or whole tone apart. The shortest step in a scale or in the interval between a pair of notes is semitone like E and F or B and C'. There are totally 12 semitones in the span of an octave in the commonly accepted diatonic scale. According the numbers of semitones used, the interval of any pair of two notes spanned within an octave may have different name, see Table 2.7.1.

Table 2.7.1 The intervals within an octave and corresponding integer ratio of frequencies

Interval name	Number of semitones	Frequency ratio (appox.)	Examples
Perfect unison	0	1/1	C-C
Minor second	1	16/15	F-E or C'-B
Major second	2	9/8 or 10/9	D-C or E-D
Minor third	3	6/5	G-E or C'-A
Major third	4	5/4	E-C
Perfect fourth	5	4/3	F-C or G-D
Augmented fourth Diminished fifth	6	45/32 64/45-7/5	B-F
Perfect fifth	7	3/2	G-C
Minor sixth	8	8/5	C'-E
Major sixth	9	5/3	A-C
Minor seventh	10	7/4 or 16/9	C'-D
Major seventh	11	15/8	B-C
Perfect octave	12	2/1	C'-C

3. Deep reading

About the frequency ratio, tuning and temperament.

The musical intervals can be expressed by the frequency ratios which follow the facts:

(1) The ratio of frequencies that defines the musical interval.

(2) Adding intervals amounts to multiplying by frequency ratios.

As we know that the frequency relation of C and C'(higher an octave) is fixed by a ratio of 1/2 accurately, similarly, C and G by 2/3. The ratio 3/2 defines the Just interval for a fifth. Alternatively, consider the sequence of notes, C-E-G. C-E is a major 3rd, while E-G is a minor 3rd. Together, they add up to the musical interval of a fifth, corresponding to C-G. We write

major third + minor third = fifth

This fact is exhibited in Fig. 2.7.5.

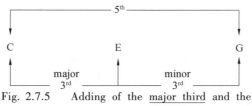

Fig. 2.7.5 Adding of the major third and the minor third to a fifth

The equation above is reflected by the following mathematical equation:

$$\frac{f_E}{f_C} \cdot \frac{f_G}{f_E} = \frac{f_G}{f_C}$$

From the viewpoint of integer ratios linked to intervals, any note defined within an octave (C-C′) in the common set of musical scale (C, D, E, F, G, A, B, C′) is perfect, since for D we have

$$\frac{f_D}{f_C} \cdot \frac{f_{C'}}{f_D} = \frac{f_{C'}}{f_C}$$

Similarly for E, F, G, A, B,···, since $(f_E/f_C)(f_{C'}/f_E) = f_{C'}/f_C$, $(f_F/f_C)(f_{C'}/f_F) = f_{C'}/f_C$, $(f_G/f_C)(f_{C'}/f_G) = f_{C'}/f_C$, $(f_A/f_C)(f_{C'}/f_A) = f_{C'}/f_C$, $(f_B/f_C)(f_{C'}/f_B) = f_{C'}/f_C$, ···

How should the frequencies associated with the notes in the diatonic scale be chosen? This choice is referred to as tuning, namely the method or process from one note to go to another one in the scale. We will discuss in detail the three most important tunings: Just tuning, Pythagorean tuning, and Equal tempered tuning.

In brief, Pythagorean tuning is the method of generating a musical scale by using descending or ascending by the fifth operation combining the octave operation.

Pythagorean tuning is based on assigning a frequency ratio of 3/2 to the fifth and treating all the remaining intervals as subservient to the fifth, as follows. We start by choosing the frequency f_{C_4}. Then, we determine the frequencies for the notes that arise as we ascend and descend in a series of fifths:

$$F_3 \to C_4 \to G_4 \to D_5 \to A_5 \to E_6 \to B_6$$
$$2/3 \quad 1 \quad 3/2 \quad (3/2)^2 \quad (3/2)^3 \quad (3/2)^4 \quad (3/2)^5$$

We will discuss only the five notes C, D, F, G and A. These notes form a pentatonic scale, which is the essential scale in many parts of the non-Western world. The process of determining the remaining frequencies that are needed to form the diatonic and chromatic scales is a mere continuation of the process we will demonstrate below.

What we want are the frequencies of notes in the same octave starting with C_4. They are determined by a process called reducing to the original octave the above frequencies, as follows: $f_{F4} = 2f_{F3}$, $f_{D4} = (1/2)f_{D5}$, $f_{A4} = (1/2)f_{A5}$. Thus, using $2(2/3) = 4/3$, $1/2(3/2)^2 = 9/8$, and $1/2(3/2)^3 = 27/16$, we obtain the pentatonic scale in Pythagorean tuning, see Fig. 2.7.6.

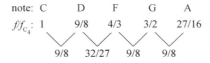

Fig. 2.7.6   Full pentatonic scale in Pythagorean tuning (The upper, middle and lower lines list the notes in the scale, frequency ratio to that of middle C, f/f_C and adjacent frequency ratio)

On the last line, we have indicated the intervals between neighboring notes. For example, for the interval between D and F, we have

$$\frac{f_F}{f_D} = \frac{\frac{f_F}{f_C}}{\frac{f_D}{f_C}} = \frac{\frac{4}{3}}{\frac{9}{8}} = \frac{4}{3} \times \frac{8}{9} = \frac{32}{27}$$

We note that the two intervals of a <u>third</u> (the <u>minor third</u> and the <u>major third</u>) each involve a ratio of rather large integers. As such, these intervals do not sound very consonant in the traditional sense.

<u>Just tuning is the method of generating a musical scale which sets a priority on ratios of small integers for the frequency ratios of musical intervals to reach the maximized consonance of a pair of notes. In the Just tuning, for the octave the frequence ratio is 2/1, for the fifth 3/2, while for the major third 5/4, which are central ones in Western classical music to generate Just notes in a musical scale.</u>

Let us lay out the notes of the <u>diatonic scale</u> with the following notation (see the upper line in Fig. 2.7.7).

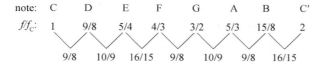

Fig. 2.7.7   Full Just diatonic scale (The upper, middle and lower lines list the notes in the scale, frequency ratio to that of central C, f/f_C and adjacent frequency ratio)

From central C, one can easier get the note E and G ascending by a <u>major third</u> and a <u>fifth</u>. We can obtain A by descending from E to A by a <u>fifth</u> and then reducing to the octave by ascending by an <u>octave</u>. This involves the product $2(2/3)(5/4) = 5/3$. We can obtain the F by descending from A by a <u>major third</u>, thus obtaining the ratio $(4/5)(5/3) = 4/3$. We can obtain the D by ascending from G by a <u>fifth</u> and then descending by an <u>octave</u>. The product is $(1/2)(3/2)(3/2) = 9/8$. Finally, we can obtain the B by ascending from E by a <u>fifth</u>. The product involved is $(3/2)(5/4) = 15/8$. Thus the full <u>Just diatonic scale</u> is obtained, see the Fig. 2.7.7. Intervals between neighboring notes are obtained by taking ratios of the corresponding pair of ratios to C. For example:

$$\frac{f_F}{f_E} = \frac{\frac{f_F}{f_C}}{\frac{f_E}{f_C}} = \frac{\frac{4}{3}}{\frac{5}{4}} = \frac{4}{3} \times \frac{4}{5} = \frac{16}{15}$$

In the similar way, any interval in an octave can also be obtained by ratios to C. For instance, we can obtain the <u>minor third</u>, 6/5, from the interval E-G as follow $f_G/f_E =$

$(f_G/f_C)/(f_E/f_C) = (5/4)/(3/2) = 6/5$.

Just intervals have the beauty of rich consonance. Just tuning tries to find a small integer ratio to express the interval of two notes to let them sound consonantly. Unfortunately, for fundamental, mathematical reasons, they are impossible to realize fully in a composition, as we will observe in the following examples.

Example 1: In the key of C, D-F is a minor third and yet has a frequency ratio of $(4/3)/(9/8) = 32/27$ instead of the standard Just minor third of 6/5. Thus, there are two different minor thirds in the scale. The ratio 32/27 is rather harsh and defeats the goal of Just tuning. We would hope that this interval could be avoided, unfortunately, mathematical analysis shows this to be impossible.

Example 2: One could choose to play the major sixth between the D and the B as a Just major sixth, that is, 5/3. But then, the interval of a major fourth between the B and the E would be $[(3/2)(3/2)](5/3) = 27/20$, rather than the Just ratio of 4/3.

The ratio between these two numbers above is $(27/20)/(4/3) = 81/80$, which can be defined a smallest interval with term of syntonic comma that the untrained ear can notice. While it is equivalent to only about one-fifth of a semitone, it is large enough to make the interval 27/16 noticeably dissonant. We will see the syntonic comma a number of times since it is ubiquitous in the mathematics of tunings.

Example 3 Considering the intervals between neighboring notes: D to G above (perfect fourth), G to E below (minor third), E to A above (perfect fourth), A to D below (perfect fifth), with Just intervals between neighboring notes, we obtain

$$f_D = f_D \times \frac{4}{3} \times \frac{5}{6} \times \frac{4}{3} \times \frac{2}{3} = \frac{80}{81} f_D \neq f_D$$

Again we see the difference in term of 81/80.

We see then that it is absolutely impossible to perform pieces fully with just intervals between all pairs of significant notes. Furthermore, the errors are significant.

We need compromise in tuning with the integer ratios, the Equal tempered tuning then is devised.

In Equal tempered tuning, the octave is divided into 12 equal semitone intervals. The gain is that the problems of Just tuning are removed. The loss is that resonances and consonances are not as strong as they are in Just tuning.

In the tuning, consonant intervals such as the fifth or the major third are harsher. The greatest gain, perhaps, is that fixed tuned instruments such as the piano do not have to be retuned for each key change. Furthermore, single compositions that have changes of key can be played in a consistent manner on fixed tuned instruments, that is, without biasing one key. For equal tempered tuning, all we need is to determine the frequency ratio corresponding to a semitone. We label it with the symbol r. Since there are 12 semitones to the octave, we must have $r^{12} = 2$, so

$$r = 2^{1/12} = 1.059\ 46$$

For comparison sake, the Just semitone is 16/15 = 1.066... and is therefore slightly larger. To compensate for this increased value within the octave, Just tuning has a whole tone interval, 10/9 = 1.111..., which is less than the equal tempered whole tone (two semitones) interval of

$$r^2 = 2^{2/12} = 2^{2/12} = 1.122\ 46$$

On the other hand, recall that Just tuning uses two different whole tones; the other Just whole tone interval of 9/8 = 1.125 is greater than r^2.

Table 2.7.2 compares the important intervals in Just (J) and Equal Tempered (ET) tuning.

Table 2.7.2 A comparisons of some intervals in Just and Equal Tempered tuning

Interval	Just (J)	Equal Tempered (ET)
Semitone	16/15 = 1.066...	$2^{1/12}$ = 1.059...
Whole tone	9/8 = 1.125, 10/9 = 1.11...	$2^{2/12}$ = 1.122...
Minor third	6/5 = 1.2	$2^{3/12} = 2^{1/4}$ = 1.189...
Major third	5/4 = 1.25	$2^{4/12} = 2^{1/3}$ = 1.26...
Fourth	4/3 = 1.333...	$2^{5/12}$ = 1.335...
Fifth	3/2 = 1.5	$2^{7/12}$ = 1.498...
Minor sixth	8/5 = 1.6	$2^{8/12} = 2^{2/3}$ = 1.587...
Major sixth	5/3 = 1.666...	$2^{9/12} = 2^{3/4}$ = 1.68...
Octave	2	2

To express intervals that are smaller than a semitone in a quantitative way, so as to reflect very small changes of pitch, the equal tempered semitone, $r = 2^{1/12}$, is divided into one hundred (100) units, called cents, as follows:

100 cents = 1 ET semitone, 1 200 cents = 1 octave

Generally, in the cents system, the interval between two frequencies, f_2 and f_1, is given by

$$C \equiv \frac{1\ 200}{\lg 2} \cdot \lg \frac{f_2}{f_1} \qquad (2.7.9)$$

where the unit of the parameter C is the number of cents in the interval.

We thus have here an explicit, mathematical expression for the musical interval as a function of the corresponding frequency ratio.

Temperament in fact refers to the various tuning systems for the subdivision of the octave. We often meet the some principal tuning systems being Pythagorean tuning, Just intonation, Equal Tempered tuning and Mean-tone tempered tuning. Accordingly, we have four temperaments, in which the first three are common used in practice.

The Pythagorean temperament uses the interval of fifth and the octave to form a circle (the circle of fifth), as shown in Fig. 2.7.8. For instance, from C one can get G above a fifth, then D, A,... or from C get F by descending a fifth, then $B_\flat$, $E_\flat$,... .

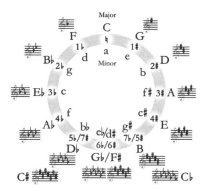

Fig. 2.7.8 The circle of fifths, the method used to generate musical sound in Pythagorean temperament

In Just temperament, every interval between two pitches corresponds to an integer ratio between their frequencies, allowing intervals varying from the highest consonance to highly dissonant. Such intervals have a stability, or purity to their sound, when played simultaneously (assuming they are played using timbres with harmonic partials). If one of those pitches is adjusted slightly to deviate from the just interval, a trained ear can detect this change by the presence of beats, which are periodical oscillations in the note's intensity.

An equal temperament is a musical temperament, or a system of tuning, in which the frequency interval between every pair of adjacent notes has the same ratio. In other words, the ratios of the frequencies of any adjacent pair of notes is the same, and, as pitch is perceived roughly as the logarithm of frequency, equal perceived "distance" from every note to its nearest neighbor. In equal temperament tunings, the generating interval is often found by dividing some larger desired interval, often the octave (ratio 2/1), into a number of smaller equal steps (equal frequency ratios between successive notes) termed as twelve-tone equal temperament, in which the octave is divided into 12 parts, all of which are equal on a logarithmic scale, with a ratio equal to the 12th root of 2 ($\sqrt[12]{2} \approx 1.059\,46$). Historically, Zaiyu Zhu, a prince in Ming Dynasty in China (Fig. 2.7.9), had first invented the relation of geometric series among musical notes and termed the twelve-tone equal temperament, which was earlier than the West almost by 100 years.

Fig. 2.7.9 The statue of Zaiyu Zhu (1536—1611) in the Museum in Henan province, China

Part 2 Topics of Experiments

In fact, in ancient Chinese music theory, a method termed as the method of "Subtracting and Adding One Third" (Fig. 2.7.10), can also be viewed as the most ancient musical temperament in the world, which was recorded in *The Master Lü's Spring and Autumn Annals*, an encyclopedic Chinese classic book compiled around 239 BC under the patronage of the Qin Dynasty Chancellor Buwei Lü.

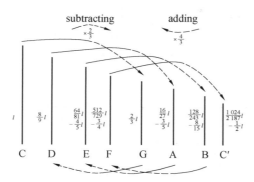

Fig. 2.7.10   The method of "Subtracting and Adding One Third"

In Lü's book, a story on constructing musical instruments by Ling Lun (the officer in charge of Music for Emperor Huangdi about 5 000 years ago) had been recorded. Ling Lun from the three parts of the "huang chung generator (a section of bamboo pipe)" rejected one part and made the "inferior generator" (hence equal to 2/3 of the huang chung generator). Next, taking three parts of the new (i.e. inferior) generator and adding one part, he made the "superior generator" (hence equal to 4/3 of the inferior generator)..., and so forth. The lengths of the bamboo pipes are based on repeated applications of the factor 2/3 and 4/3 on the basic length of the huang chung generator which involves alternately. More explicitly, if the initial tone of the huang chung is C in the key of C, after first treatment, one can obtain G a fifth above C. After second treatment, from G one can get D a fourth apart, then A, E, B,.... From this temperament, from one tone, all 12 tones nearby within an octave can be generated. Actually it is more convenient method than Pythagorean one to generate the musical scale. The fact that the relation of frequency ratio can be expressed by a ratio of small integers is obvious and natural.

About the consonance (harmony) or dissonance of combined sounds.

Consonance and dissonance are obviously related to harmonics, or more exactly, the spectrum of harmonics, so let's look at how they are related.

Without loss of generality, consider the fifth, the dyad, a form of chord, we know that it is one of the most harmonic combinations in music. Why? For an answer we have to look at the spectrum of harmonics of the two tones. Let's begin with the first tone (see first row in Fig. 2.7.11). Its fundamental is f_1 and its overtones are $2f_1$, $3f_1$,... . For the tone of perfect fifth we have the spectrum shown in second row. Bringing them together, we get the spectrum shown in third row. We see immediately that several tones match up; furthermore, tones that

75

do not match are a considerable distance from one another. Considerable separation of unmatched tones is important because tones close together will create beats, which are a problem.

We can also do the same type of analysis for the fourth, the third, and the other intervals. (On the piano the fifth is the interval C-G, the fourth is C-F, and the major third is C-E.) If we do this for all types of intervals and arrange the intervals in order of greatest to least consonance (i.e., dissonance), we get Table 2.7.3, showing the greatest consonance at the top and increasing dissonance as you go down the column.

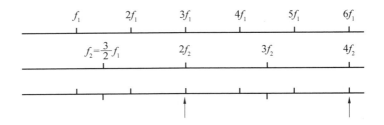

Fig. 2.7.11   Spectra of harmonics for the first tone, the perfect fifth and combined spectrum for the first tone and the perfect fifth

Table 2.7.3 **Intervals from greatest consonance to great dissonance**

Interval	Notes (key of C major)	
Octave	C-C′	Greatest consonance
Fifth	C-G	
Fourth	C-F	
Major third	C-E	
Major sixth	C-A	↓
Minor third	C-E♭	
Minor sixth	C-A♭	
Whole tone	C-D	
Half tone	C-D♭	Greatest dissonance

4. Experiment contents

(1) Open the interface programmed by LabVIEW for analyzing acoustic signal in the computer and learn about the parameter setting for signal sampling and manipulation. Sample the noise and musical signal by using a microphone inserted to the computer. Differentiate these two kinds of signals in the characteristics in time domain and frequency domain. Understand the concepts on frequency component, main frequency, fundamental frequency, overtone and harmonics (second, third harmonic...). Understand the link between a certain

component and the vibrational mode. Take the pictures.

(2) Differentiate some musical sounds (three tones at least) at the same pitch (key note) from various sources such as musical instruments, human throat to learn more about difference in timbre and determine the same points of the tones. Record the data of their frequencies (fill in the given table) and take the pictures accordingly.

(3) Study the frequency evolution of musical notes in a musical scale within the octave (Do, Re, Mi, Fa, Sol,...) using keyboard instrument (electronic synthesizer) by fixing a timbre (by setting number on the timbre selection keyboard) and record the corresponding fundamental frequency and main frequency by filling in the given table. Perform the calculation for the frequency ratios accordingly.

(4) Master the usage of digital store oscilloscope in display and measurement of the signal period and the frequency. Repeat the comparison of musical tones in the different timbres by filling in the given table. Understand the correspondence with fundamental and main frequency.

(5) Study the combination of musical sounds such as chord (dyad, triad) and understand the nature of consonance of a chord like the major triad (Do + Mi + Sol). (selective topics)

(6) Study the method of "Subtracting and Adding One Third" to generate a notes in musical scale using test tubes and the water and record the pitch of each produced tones and compare the relation between. (selective topics)

(7) Study some interesting methods to generate musical sounds such as using the effect of Helmholtz cavity, whistle effect... (selective topics)

5. Cautions

When you use water in the experiment, you should avoid the water sprinkling to the electrical instrument.

6. Discussion

(1) Imagine yourself seated in a concert hall, anticipating the beginning of a symphony orchestra performance. The musicians are all seated. The concertmaster rises and calls to the oboist to sound the "A", which will be the standard pitch that all others will use to tune their instruments and has been tuned to 440 Hz. Next, the strings tune their A-strings to the oboe's "A". Following that, the strings tune their other three strings accordingly. For example, the violins tune the "E", which is a musical "perfect fifth" above the "A". Usually, a violinist play "E" simultaneously with the "A". And strives for the maximal beauty of a resonance (consonance or harmony) between the two strings. The violinist continues in a similar manner with the "D" a fifth below the "A" and then the "G" a perfect fifth below the "D". Why they do so? Try to explain it!

(2) What is the frequency ratio between the adjacent strings for a well-tuned guitar? Is it the same with that of the well-tuned violin?

(3) Do you agree with the opinion that the Piano may be viewed as the King of Music while the violin — the Queen? Tell your reasons, please!

Experimental Report

Exploring the Musical Sounds

Name: Student ID No.: Score:

Date: Partner: Teacher's signature:

1. Purpose

2. The experimental data and processing

(1) Understanding the timbre (Table 2.7.4 – 2.7.5).

Table 2.7.4 Comparison of tones from different sources keeping the same pitch using LabVIEW-based sound analyzing interface

Tones from different sources	Waveform	Frequency spectrum	Fundamental frequency /Hz	Main frequency /Hz
Tone 1 ( )				
Tone 2 ( )				
Tone 3 ( )				

Table 2.7.5 Comparison of the tones from different sources keeping the same pitch using digital storage oscilloscope

Tones from different sources	Waveform	Period /ms	Frequency /Hz
Tone 1 ( )			
Tone 2 ( )			
Tone 3 ( )			

Conclusion about the difference and the same place of tones:

(2) Musical scale and interval relation (within an octave) (Table 2.7.6).

Table 2.7.6 Evolution of frequency of musical notes within an octave in key of <u>C</u> and determination of frequency ratio keeping the same timbre and using LabVIEW-based sound analyzing interface

Notes	Fundamental frequency $f_{i0}/$Hz	Main frequency $f_{im0}/$Hz	Frequency ratio Experimental f_{i0}/f_{Do0}	Theoretically predicted by Pythagorean temperament	Just temperament	12-equal temperament	Method of "Subtracting and Adding One Third"
C			1	1	1	1	1
C#							
D							$\frac{3}{2} \cdot \frac{3}{4} = \frac{9}{8}$
D#					$\frac{6}{5}$		
E					$\frac{5}{4}$		
F					$\frac{4}{3}$		
G♭							
G				$\frac{3}{2}$	$\frac{3}{2}$		$\frac{3}{2}$
A♭							
A							
B♭							
B							
C'					$\frac{2}{1}$	$\frac{2}{1}$	

Conclusion about the frequency ratio:

The experiment data mainly obey the law predicted by _____ temperament!

(3) Understanding the consonance or harmony of a chord (selective topics).

Record the change when separately playing the note <u>C, E, G</u> and playing them together (a chord), take the picture about the timbre in time and frequency domain (Table 2.7.7).

Table 2.7.7   Comparison of the change by playing the chord and separately play the composing notes by keeping the same timbre of the notes

Tones composing a chord (keeping the same timbre)	Spectrum, distribution of frequency components	Fundamental frequency /Hz	Periodicity yes√ no ×	Coincidence of frequency components yes √ no ×
C				—
E				—
G				—
dyad, C + E				
dyad, C + G				
dyad, E + G				
major triad, C + E + G				

Conclusion about the understanding the consonance:

(4) Understand the method of "Subtracting and Adding One Third" to generate a notes in musical scale using test tubes and the water (selective topics).

(5) Interesting experiments for generating musical sounds (Helmholtz cavity effect, selective topics) (Fig. 2.7.12).

Prove the relation between sound frequency generated from a cavity by blowing a bottle from the neck and its air volume inside as

$$f = \frac{v}{2\pi}\sqrt{\frac{S}{VL}}$$

where V is the volume of the air inside the container; v is the velocity of the air jet; S, L denotes the cross sectional area and the length of a bottle neck, respectively.

Change the volume V by filling and pouring water way and record the fundamental frequency f by performing filling in a self-designed table. Draw the dependence of frequency on Volume, namely the $f - V$ curve in a piece of graph paper and prove the relation.

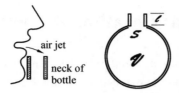

Fig. 2.7.12   Blowing bottle from the neck

3. Discussion

2.8 Forced Vibration of Pohl's Pendulum

Resonance, as a phenomenon widely existing in mechanics, electricity, acoustics, optics and structure of substance, is extensively studied and found wide applications in some fields such as machinery manufacture, construction industry, electro-acoustic device design, equipment of nuclear magnetic resonance, and so on. In this experiment, a Pohl pendulum is employed to investigate the interaction between the active stimulation and the negative motion of a system. The frequency-dependence of the vibration amplitude and phase in the negative system will be studied. The damping (attenuation) coefficient of the system will be determined. The stroboscopic method is utilized to measure the transient phase of the system.

1. Purpose

(1) To study the resonance caused by interaction of an active stimulator and a negative system;

(2) To study the damping effect on vibration and determine the damping coefficient of the negative system;

(3) To measure the amplitude-frequency characteristics and the phase-frequency characteristics of a forced vibration system;

(4) To master the stroboscopic method to measure the transient phase of the system.

2. Principle

(1) The vibration of Pohl pendulum.

The Pohl pendulum is such a mechanical negative system, in which a metal circular disk may swing (rotate back and forth) on its axis in a damping environment. When a force (periodic stimulus) is applied to the system, it will reciprocate in response to the force. The motion belongs to a kind of forced vibration.

In the vibration, a damping torque (moment) M_1, an elastic restoring torque M_2 and the periodic stimulating torque M_3 mainly act on the pendulum. In the low speed approximation, damping torque is directly proportional to the rotational speed, namely $M_1 = -\gamma d\theta/dt$, where $d\vartheta/dt$ is the rotational speed, γ is the proportional coefficient, namely the damping moment coefficient. The elastic restoring torque can be expressed by $M_2 = -k\theta$, where θ is the rotational angle (angular displacement), k is the elastic restoring torque coefficient, that means the lager the angle the lager the restoring torque. The periodic stimulation can be expressed as $M_3 = M_0 \cos \omega t$, where M_0 is the amplitude of stimulation, ω is its angular frequency. If the inertia of the pendulum is I, we have the kinetic equation to describe the movement of the pendulum:

$$I\frac{d^2\theta}{dt^2} = -\gamma \frac{d\theta}{dt} - k\theta + M_0 \cos \omega t \qquad (2.8.1)$$

or

$$\frac{d^2\theta}{dt^2} + 2\beta\frac{d\theta}{dt} + \omega_0^2\theta = h\cos\omega t \qquad (2.8.2)$$

where β is damping coefficient, $2\beta = \gamma/I$; ω_0 is the inherent frequency determined by $\omega_0^2 = k/I$. Here $h = M_0/I$.

①Damped vibration. If there is no stimulation on the system, the kinetic equation becomes

$$\frac{d^2\theta}{dt^2} + 2\beta\frac{d\theta}{dt} + \omega_0^2\theta = 0 \qquad (2.8.3)$$

It is the typical second-order homogeneous differential equation with constant coefficients, whose solution is

$$\theta = \theta_0 e^{-\beta t}\cos(\omega t + \varphi) \qquad (2.8.4)$$

where θ_0, ω and φ denote the initial angle, the vibrational angular frequency and phase, respectively. Notice that

$$\omega = \frac{2\pi}{T} = \sqrt{\omega_0^2 - \beta^2} \qquad (2.8.5)$$

where T is the vibrational period.

The solution suggests that the vibration is damped with a factor β. After nT the amplitude becomes $\theta_{0n} = \theta_0 e^{-\beta nT}$, so we have a relation of

$$\beta = \frac{1}{nT}\ln\frac{\theta_0}{\theta_{0n}} \qquad (2.8.6)$$

which can be used to determine the damping coefficient by measuring the initial amplitude and the final amplitude after the time of n periods, namely nT.

②Forced vibration. When the periodic stimulus exists, the kinetic equation becomes the second-order nonhomogeneous differential equation with constant coefficients, whose solution can be expressed as

$$\theta = \theta_0 e^{-\beta t}\cos(\sqrt{\omega_0^2 - \beta^2}\,t + \varphi_0) + \theta_1\cos(\omega t + \alpha) \qquad (2.8.7)$$

It should be pointed out that from the expression above the forced vibration is a combination of two kinds of vibrations: the damped one and the stable one. With increase of time the former tends to zero because of the factor of $e^{-\beta t}$, which becomes negligible after an enough interval of time, whereas the latter — a steady state of vibration with a stable period and an amplitude remain. The steady-state solution with the form of $\theta = \theta_1\cos(\omega t + \alpha)$ is often used to characterize the final state of a forced vibration system. It is clear that then the frequency or the period of a negative system is the same with that of stimulus, which becomes an indicator representing the negative system gets into the steady state and is in coordination with the stimulation.

In fact, for a forced vibration in the steady state both the amplitude and the phase (with respect to the phase of the stimulation) are the function of the stimulating frequency, namely theyare frequency-dependent. We call the dependence of the amplitude and the phase on the stimulating frequency the amplitude-frequency characteristics and phase-frequency characteristics. Theoretically the amplitude and the phase can be expressed by

$$\theta_1 = \frac{h}{\sqrt{(\omega_0^2 - \omega^2)^2 + 4\beta^2\omega^2}} \quad (2.8.8)$$

and

$$\alpha = \arctan\frac{-2\beta\omega}{\omega_0^2 - \omega^2}, \quad \omega \leqslant \omega_0 \quad (2.8.9)$$

$$\alpha = -\tau + \arctan\frac{-2\beta\omega}{\omega_0^2 - \omega^2}, \quad \omega > \omega_0 \quad (2.8.10)$$

From Eq. (2.8.8) one can see that when ω approaches zero the amplitude of forced vibration θ_1 tends to that of external stimulating torque $h/\omega_0^2 = \theta_0$ (since $h = M_0/I = k\theta_0/I = \omega_0^2\theta_0$). First with increase of ω, θ_1 increases gradually. When $\omega = \sqrt{\omega_0^2 - 2\beta^2} = \omega_R$, the system gets into the resonance state, θ_1 reaches the maximum θ_R with

$$\theta_R = \frac{h}{2\beta\sqrt{\omega_0^2 - \beta^2}} \quad (2.8.11)$$

Here ω_R is called the resonance frequency. When ω continues to increase ($\omega > \omega_R$), θ_1 decreases monotonically. When ω reaches ∞, θ_1 becomes zero. From Eq. (2.8.11), one can find that the smaller the damping coefficient the larger the maximum amplitude of steady state, namely the higher and sharper the resonance curve. Fig. 2.8.1 compares several resonance curves with different damping coefficients, from which one may feel the role of damping coefficient.

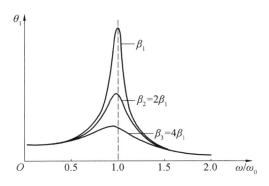

Fig. 2.8.1 Amplitude-frequency characteristics

From Eq. (2.8.9) and Eq. (2.8.10), one can see that when $0 \leqslant \omega \leqslant \omega_0$, $0 \leqslant \alpha \leqslant -\pi/2$, that means forced vibration obviously lags behind the stimulus in phase. At the resonance, $\alpha = -\pi/2$. When $\omega > \omega_0$, the phase lags more than $\pi/2$. When $\omega \gg \omega_0$, α approaches to $-\pi$. Known ω_0 and β the phase lag at any frequency ω can be determined. Fig. 2.8.2 shows two curves of the phase-frequency characteristics with different damping coefficient. In the experiment, the phase α can be measured by utilizing the stroboscopic method, which is often used in studying a moving object with the periodic stroboscope. The photogate and the control circuit of the system guarantees the flash is turned on when the metal disc is instantly located at the status of zero phase.

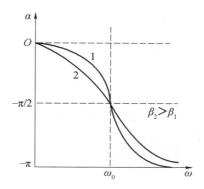

Fig. 2.8.2 Phase-frequency characteristics

3. Instruments

The structure of Pohl pendulum used in the experiment is shown in Fig. 2.8.3. As the negative system, the metal wheel pendulum (metal disk) may rotate on a fixed axis and connected to a volute spring. The damping can be controlled electromagnetically by the current flowing through a damping coil. As the active stimulus, a motor, the rotation speed of which can be adjusted by a variable resistor, can exert the influence on the pendulum through the pitman structure.

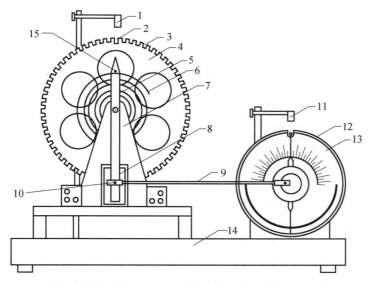

Fig. 2.8.3 The structure of Pohl pendulum (resonator)

1—photogate; 2—long groove; 3—short groove (tooth); 4—metal wheel pendulum; 5—rocking bar; 6—volute spring (ribbon spring in windup clock); 7—support bracket; 8—damping coil; 9—connecting rod (pitman); 10—adjusting screw; 11—photogate; 12—disk with angular scales; 13—polymer turning disc; 14—pedestal; 15—holding screw

In the experiment, the mechanical part of Pohl pendulum is linked to a control box. The picture of control panel is shown in Fig. 2.8.4, where the "up" and "down" keys are used to select the some subject while the "left" and "right" keys to change the operation status of the pendulum. The enter key ("确认") is used to affirm the selected function. The reset key ("复位") is used to recover the initial state of the system. The rotary knob ("强迫力周期") can change the stimulating period.

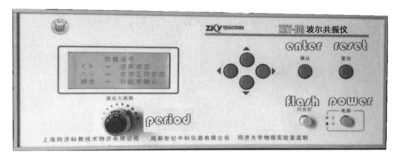

Fig. 2.8.4　The control panel of Pohl pendulum

4. Experiment contents

(1) Observe the free vibration and roughly determine the inherent period and frequency of the vibration in the mode of "free vibration(自由振动)".

(2) Record the damping behaviors ($\theta - t$) under different electromagnetic damping in the mode of "vibration with resistance(阻尼振动)".

Note: choose the position of resistance "1" and "3" for the electromagnetic damping. The initial angle mustn't excess 120°, the measurement must be performed under the "10 periods" mode with the switch of the electromotor on the state of "off" and the connecting rod at horizontal level.

Record the change of vibrational angel and the corresponding time as fast as possible (practice please before starting record!).

(3) Measure the amplitude-frequency characteristics and phase-frequency characteristics.

The measurement must be performed under the "1 periods" mode (single mode) with choice of the position of resistance "1" and "3" for the electromagnetic damping and the state of "on" for switch of the electromotor. The rotation speed of the motor can be changed by the knob of period adjustor. After a while with a fixed resistor the system may get in the steady state: the display for period and amplitude become constants, record them, then switch on the flash to let it the irradiate the polymer turning disc and read the temporal position of the white line (with fluorescent powder inside) on it. Change the speed of the motor with resistor, repeat the procedure again.

Note: it is not allowed that the forced vibration is performed without electromagnetic damping (at least at "1" position) to avoid the damage of the mechanical system. It is suggested that first observe the phenomenon of resonance (the amplitude reach a maximum at a certain angular frequency of motor and the angular position on the motor's disk irradiated by the flash indicated with a white line is near 90°), then measure the characteristics by scanning the frequency (suggested from low to higher or vice versa). In order to prettily draw the curves the data near the resonance must be recorded denser than that for far from resonance (in the "off-resonance" range).

5. Cautions

In the experiment, it is not allowed to prevent or block the rotation of the turning discs by halting the metal or polymer discs. It may cause a motor burnout and the damage of the mechanical system.

Don't touch the photogates when the system is in operation.

In order to save the lifetime of flash turn flashlight on when the system gets in the steady state.

6. Discussion

(1) How to get α by reading the temporal position of while line on the polymer disc.

(2) Why the white line on disc has two steady positions? How to solve this problem?

Experimental Report

Forced Vibration of Pohl's Pendulum

Name:　　　　　　Student ID No.:　　　　　　Score:

Date:　　　　　　Partner:　　　　　　Teacher's signature:

1. Purpose

2. The experimental data

(1) Measurement of period and frequency in the mode of free vibration.

$T_0 = $ _____ (s); $\omega_0 = $ _____ (Hz).

(2) Record of damping behaviors ($\theta - t$) in the mode of damped vibration (Table 2.8.1, Table 2.8.2)

Table 2.8.1　The amplitudes in every period at damping position "1" with n periods $nT_1 = $ _____ (s)

Periods T/s	1st	2nd	3rd	4th	5th
Amplitude θ/(°)					
Periods T/s	6th	7th	8th	9th	10th
Amplitude θ/(°)					

Table 2.8.2　The amplitudes in every period at damping position "3" with n periods $nT_3 = $ _____ (s)

Periods T/s	1st	2nd	3rd	4th	5th
Amplitude θ/(°)					
Periods T/s	6th	7th	8th	9th	10th
Amplitude θ/(°)					

(3) Measurement of amplitude-frequency and phase-frequency characteristics. Observe

the resonance phenomenon first and distribute the data adequately (Table 2.8.3, Table 2.8.4).

Table 2.8.3   Data for $\theta_1 - \omega$ and $\alpha - \omega$ at damping position "1"

No.	1	2	3	4	5	6	7	8
T/s								
ω/Hz								
θ_1/(°)								
α/(°)								
No.	9	10	11	12	13	14	15	16
T/s								
ω/Hz								
θ_1/(°)								
α/(°)								

Resonance stimulating period $T_r =$ _____ (s).
Resonance frequency $\omega_r = 1/T_r =$ _____ (Hz).

Table 2.8.4   Data for $\theta_1 - \omega$ and $\alpha - \omega$ at damping position "3"

No.	1	2	3	4	5	6	7	8
T/s								
ω/Hz								
θ_1/(°)								
α/(°)								
No.	9	10	11	12	13	14	15	16
T/s								
ω/Hz								
θ_1/(°)								
α/(°)								

Resonance stimulating period $T_r =$ _____ (s).
Resonance frequency $\omega_r = 1/T_r =$ _____ (Hz).

3. Data processing

(1) Calculate the decay rate (damping coefficient).

① at damping position "1".

$$\beta_1 = \frac{1}{nT_1} \ln \frac{\theta_0}{\theta_{0n}} = \frac{1}{10T} \ln \frac{\theta_0}{\theta_{0n}} =$$

② at damping position "3".

$$\beta_3 = \frac{1}{nT_3} \ln \frac{\theta_0}{\theta_{0n}} = \frac{1}{10T} \ln \frac{\theta_0}{\theta_{0n}} =$$

(2) Draw the curves for $\theta_1 - \omega$ and $\alpha - \omega$ at different decay rates in same coordinate paper below.

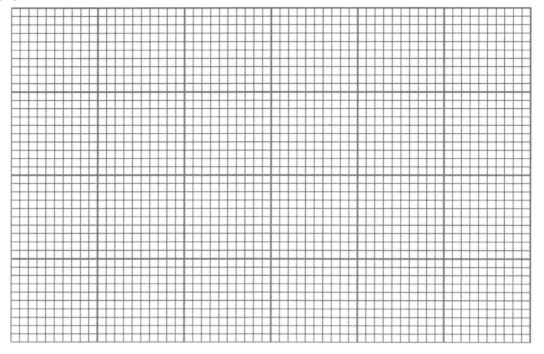

4. Discussion

2.9 Thermocouple and Its Application

Thermocouple is an important device to measure unknown ambient temperature. It has the advantages of wide measuring range, high precision, high sensitivity, small heat capacity and so on. As a direct contact temperature sensor, thermocouple is widely used in various fields.

1. Purpose

(1) To master the calibration method for the copper-constantan thermocouple;
(2) To master the usage of a thermocouple to measure the temperature difference;
(3) To study the cooling characteristics of Tin.

2. Principle

(1) The principle of thermocouple for measuring temperature.

Usually a thermocouple consists of two wires of two different materials that are joined at each end. When these two junctions are kept in different temperature environment (called the hot end and the cold end), a small electric current or a voltage drop is induced, as shown in Fig. 2.9.1. The voltage drop depends on the temperature difference between the two ends. This phenomenon is known as the <u>Seebeck Effect</u>.

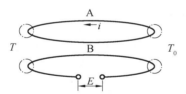

Fig. 2.9.1   Thermocouple: the current induction and voltage drop measurement

It is important to note that a thermocouple does not directly measure the target temperature but rather the temperature difference between the two junctions. In order to facilitate the measurement, one junction must be maintained at a known temperature. This junction is commonly called the reference junction or reference end. The other junction, which is normally placed in contact with the body or environment of unknown temperature, is called the measured junction.

The value of the voltage drop (or the thermal EMF, thermal electromotive force) can be correlated to the temperature difference by

$$E = \alpha(T - T_0) + \beta(T - T_0)^2 + \gamma(T - T_0)^3 + \cdots \quad (2.9.1)$$

where T and T_0 corresponds to the target temperature and the reference temperature, respectively. α, β, γ represents the linear, quadratic and cubic <u>Seebeck coefficient</u>, respectively. In fact, the linear term (the first one on the right side) is the dominant one for many combinations of different kinds of metal wires. In this experiment, we will use the thermocouple composed of copper-constantan combination. For such a kind of thermocouple, the higher-order terms can be ignored in the expression if the measured temperature ranges from $-100\ ℃$ to $400\ ℃$. That means one may reasonably view the thermocouple the linear one, in which the corresponding voltage drop E is directly proportional to the temperature difference:

$$E = \alpha(T - T_0) \tag{2.9.2}$$

(2) Calibration of thermocouple.

A thermocouple constructed or fabricated must experience a calibration before it is used in applications. Actually, the calibrating thermocouple is the procedure to check the linearity in relation between thermal EMF and temperature difference therefore to determine the Seebeck coefficient α.

Fixing the reference temperature T_0 and giving several known target temperatures T, one can measure the corresponding E. By using these data, an $E - T$ dependence — the calibration curve (usually a straight line as shown in Fig. 2.9.2) can be drawn, from which the Seebeck coefficients can be determined. In real case, the boiling point of water (corresponding temperature $100\ ℃$) and the melting point of Tin (corresponding temperature $232\ ℃$) is specially selected as two known target temperatures, through which a straight line can be obtained.

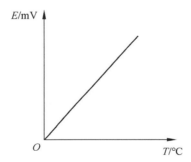

Fig. 2.9.2 Calibration curve

It should be noted that with the assistance of the straight line, the linear Seebeck coefficient α is easily obtained, and the thermal EMF difference can be easily determined by reading the temperature difference; on the other hand, the temperature difference can be well determined by reading the corresponding thermal EMF difference.

3. Instruments

(1) Heating system.

In this experiment two heaters are employed, as shown in Fig. 2.9.3, the one is for heating water and the other one is for heating Tin. We use the two pairs of thermocouples, the two hot junctions are separately placed in two containers for hot water and Tin on the heaters, while the two cold junctions are immersed in the mixture of ice-water as a common cold end.

Fig. 2.9.3　Heating system with two heaters and common cold end

(2) Digital multimeter.

Digital multimeter is often used to measure the voltage, current, resistance, etc. Fig. 2.9.4 gives the front panel of a digital multimeter used in the experiment, from which one can see that there are mainly three parts — the input terminals, the primary and secondary displays, and some push buttons for controlling and selecting a certain function. Note that the [SHIFT] button is used to enable the secondary function of certain function keys that with blue symbols printed above. The SHIFT LED will be on after the [SHIFT] button pressed. At this time, only the buttons with blue symbols are activated. To release SHIFT function, press [SHIFT] again. For example, to select "DC mV" function, press [SHIFT], and then press [DCV] ([DC mV]). In this experiment, the "DC mV" function is selected when we want to measure the voltage.

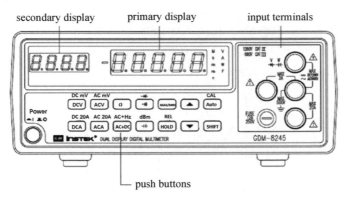

Fig. 2.9.4　Digital multimeter

(3) Stopwatch.

The stopwatch is used to confirm the time interval as shown in Fig. 2.9.5.

Fig. 2.9.5   Multi-functional temperature controller

4. Experiment contents

(1) Measure EMF for boiling water.

Fill the cup with water up to 2/3 of its capacity. Heat the water to the boiling point. Measure the EMF for boiling water by repeating 5 times, then calculate the average EMF and complete Table 2.9.1 in the experimental report.

(2) Measure EMF of Tin during a cooling process.

Heat the Tin to melting until the EMF reading is higher than 13.0 mV, then stop heating. Record EMF for Tin from 13.0 mV down to 6.0 mV every 30 s, and complete Table 2.9.2 in the report.

(3) Plot cooling curve, calibration curve.

Plot the cooling curve for Tin (like the curve in Fig. 2.9.6), find the melting point of Tin, confirm the phase transition stage ("flat stage") and record the corresponding EMF. Complete Table 2.9.3 in the report. Plot the calibration curve (one should search a straight line to "best fit" the data, like the straight line in Fig. 2.9.6).

(4) Calculate the cooling rate.

Arbitrarily choose a point A in solid phase (after the "flat stage"). Draw a tangent line through point A, choose point B and point C on the tangent line from either side, Read the corresponding coordinates $B(t_B, E_B)$ and $C(t_C, E_C)$.

Through B and C respectively, draw two lines parallel to x-axis, which intersect the calibration curve at point B' and C', read the corresponding coordinates $B'(T_{B'}, E_{B'})$ and $C'(T_{C'}, E_{C'})$, as shown in Fig. 2.9.6.

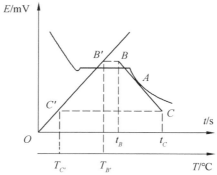

Fig. 2.9.6   Calibration curve and cooling curve

Then, we can calculate the cooling rate K_A with

$$K_A = \left|\frac{\Delta T}{\Delta t}\right| = \left|\frac{T_{B'} - T_{C'}}{t_B - t_C}\right| \qquad (2.9.3)$$

We can also calculate linear Seebeck coefficient α with

$$\alpha = \frac{E_{B'} - E_{C'}}{T_{B'} - T_{C'}} \qquad (2.9.4)$$

5. Discussion

When measuring the temperature of an object with a thermocouple, does the temperature at the reference end have to be 0 ℃, or will it have any effect?

Experimental Report

Thermocouple and Its Application

Name: Student ID No.:   Score:

Date: Partner: Teacher's signature:

1. Purpose

2. The experimental data and processing

(1) Measure the EMF for boiling water (Table 2.9.1).

Table 2.9.1   Measurement of the EMF for boiling water

Times	1	2	3	4	5	Average
E/mV						

(2) Measure the EMF for cooling melted Tin (Table 2.9.2).

Table 2.9.2   Measurement of the EMF for cooling melted Tin

t	0	30″	1′00″	1′30″	2′00″	2′30″	3′00″	3′30″	4′00″
E/mV	13.00								
t	4′30″	5′00″	5′30″	6′00″	6′30″	7′00″	7′30″	8′00″	8′30″
E/mV									
t	9′00″	9′30″	10′00″	10′30″	11′00″	11′30″	12′00″	12′30″	13′00″
E/mV									
t	13′30″	14′00″	14′30″	15′00″	15′30″	16′00″	16′30″	17′00″	17′30″
E/mV									
t	18′00″	18′30″	19′00″	19′30″	20′00″	20′30″	21′00″	21′30″	22′00″
E/mV									

(3) Calibrate the thermocouple (Table 2.9.3).

Table 2.9.3   Calibration of a thermocouple

T/°C	50	100	140	200	232
E/mV	2.035		6.204	9.288	

(4) Draw a cooling curve and a calibration curve in a graph paper, as shown in Fig. 2.9.6.

Part 2　Topics of Experiments

①Find points A, B, C, B', C', then write down coordinates for $B(t_B, E_B)$, $C(t_C, E_C)$, $B'(T_{B'}, E_{B'})$ and $C'(T_{C'}, E_{C'})$.

②Calculate the cooling rate K_A with Eq.(2.9.3) and linear Seebeck coefficient α with Eq.(2.9.4). Don't forget Units.

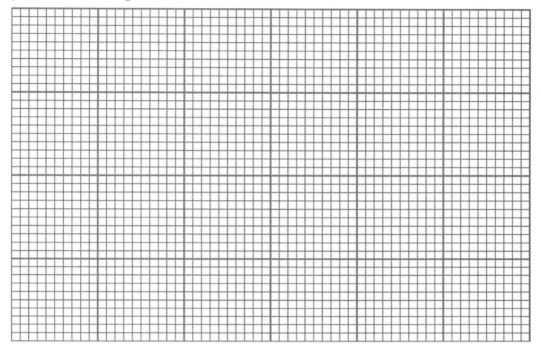

3. Discussion

2.10 Wheatstone Electrical Bridge and Its Application

1. Purpose

(1) To be familiar with the configuration and measurement principle of Wheatstone bridge;

(2) To practice the connection of the circuit and get the method for measuring an unknown resistance using Wheatstone bridge;

(3) To learn the operation of the box-bridge (integrated bridge).

2. Principle

(1) Basic principle.

Fig. 2.10.1 shows the schematic principle of a Wheatstone bridge, where R_x denotes the unknown resistance to be measured, R_1, R_2 and R_3 represent the resistors with known resistances. R_3 is adjustable. V_g and E denote the galvanometer and power supply, respectively. The balance indicator (i.e. the galvanometer) can also be replesed with a voltage meter.

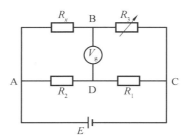

Fig. 2.10.1   Principle of Wheatstone bridge

If the ratio of R_2 and R_1 (R_2/R_1) is equal to that of R_x and R_3 (R_x/R_3), the electric potentials at B and D are equal and the galvanometer (or voltage meter) will point to zero. In such a situation, we say the bridge is in balance and we have

$$\frac{R_2}{R_1} = \frac{R_x}{R_3} \qquad (2.10.1)$$

So the unknown resistance can be measured via three known resistances by

$$R_x = R_3 \frac{R_2}{R_1} \qquad (2.10.2)$$

Usually one can fix the ratio of R_2 and R_1, and adjust the R_3 to let the bridge balanced. The R_2-to-R_1 and R_x-to-R_3 arms are named as the ratio arm and the measured arm, respectively. For convenience, the ratio $N = R_2/R_1$ are specially chosen as a ratio of two integers, for example, $N = 100, 10, 1, 0.1, 0.01,...$ as seen in the experiment using the box-bridge.

In practice, in order to get enough effective data of R_x, one should reasonably choose the resistances of R_2 and R_1 or the N in the box-bridge.

(2) Bridge sensitivity.

When a galvanometer is used as the balance indicator, the bridge sensitivity can be expressed by

$$S = \frac{\Delta G}{\Delta R_3} R_3 \qquad (2.10.3)$$

where ΔG means the change in galvanometer reading value (in Grid), induced by a small change ΔR_3 (in Ω) in R_3.

When a voltage meter is used as the balance indicator, the bridge sensitivity can be expressed by

$$S = \frac{\Delta U}{\Delta R_3} R_3 \qquad (2.10.4)$$

where ΔU means the change in voltage meter reading value (in Volt), induced by a small change ΔR_3 (in Ω) in R_3.

The sensitivity may decide the measurement accuracy of the bridge. When the system is not balance, there must be a electric potential difference between the two midpoints ("B" and "D"), which leads to a nonzero reading value on the galvanometer (or voltage meter). The current direction depends on the sign of electric potential drop, and the magnitude depends on the electric potential difference that is related with the internal resistance of the galvanometer (or voltage meter) and the voltage of the power supply. Since the internal resistance of the galvanometer is very small, we need special ways to protect it. It is suggested that firstly decrease the bridge sensitivity by using lower output of the power supply E (Fig. 2.10.2). When the circuit approaches balance, we can increase the sensitivity by turning up the output of E or directly removing the protecting resistor connected with V_g. S is a switch built in the galvanometer for protection. This switch is always at OFF state, and it will be ON only if you press the "CONNECT" button.

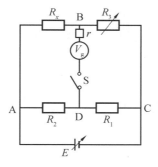

Fig. 2.10.2 The circuit of Wheatstone bridge with protection resistor (r) and switch (S)

(3) Box-bridge.

Fig. 2.10.3 gives the appearance of a box-bridge and its circuit principle. The setup can be used independently because of having built-in battery and galvanometer although it may externally connect a power supply and a galvanometer through poles $B_A^+, B_A^-, G_A^+, G_A^-$. The measured resistor can be inserted between the two poles X_1 and X_2. The knob on the upper left can be used to choose the ratio N, the four knobs on the upper right are used to realize balance of the bridge by observing the status of the galvanometer located on the lower left. The buttons B, G_0 and G_1 are used to control the connection of battery and galvanometers.

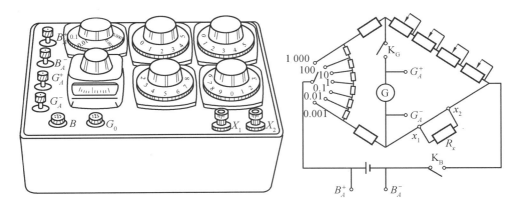

Fig. 2.10.3 Appearance of a box-bridge and its circuit principle

3. Instruments

Three unknown resistors, three resistance boxes, power supply, box-bridge, digital voltage meter (DVM), wires.

4. Experiment contents

(1) Build a Wheatstone bridge (the homemade bridge).

(2) Measure the resistances of three unknown resistors using the homemade Wheatstone bridge.

(3) Measure the resistance of the unknown resistors using the box-bridge, choose appropriate N to guarantee the calculation accuracy of R_x.

Notes: All the results should have at least 4 significant figures and all resistance boxes should use at least 100 Ω.

5. Measurement and data processing

(1) All the measured data for homemade Wheatstone bridge should be filled in Table 2.10.1 in the report, where

$$R_x = R_3 \frac{R_2}{R_1} \quad (2.10.5)$$

Note: $N = R_2/R_1$ should be the integer power of 10 such as 100, 10, 1, 0.1, 0.01, etc. All resistance boxes should be adjusted to no less than 100 Ω to guarantee the precision. Voltage of the power supply should be lower than 6 V. The suggested voltage at first is 1 V. After you build and balance the bridge, increase the voltage to 6 V and balance the bridge again.

(2) All the measured data for box-bridge should be filled in Table 2.10.2 in the report. Calculate the sensitivity of box-bridge with the equation

$$S = \frac{\Delta G}{\Delta R} R \qquad (2.10.6)$$

6. Discussion

(1) In the experiment, if a wire between *AB* or *AD* or *BD* is broken, what will happen?

(2) How to improve the sensitivity of Wheatstone bridge? Give 2 methods please.

Experimental Report

Wheatstone Electrical Bridge and Its Application

Name: Student ID No.:   Score:

Date: Partner: Teacher's signature:

1. Purpose

2. Measurement and data recording (Table 2.10.1 – 2.10.2)

Table 2.10.1   Measured resistances by using of the self-assembled bridge

	R_1/Ω	R_2/Ω	R_3/Ω	R_2/R_1	R_x/Ω
R_{x_1}/Ω					
R_{x_2}/Ω					
R_{x_3}/Ω					

Table 2.10.2   Measured resistances by using of the box-bridge

	R/Ω	N	R_x/Ω	$\Delta R/1\,\Omega$	ΔG	S
R_{x_1}/Ω						
R_{x_2}/Ω						
R_{x_3}/Ω						

3. Discussion

2.11 Measuring the Temperature Coefficient of Resistance Using Unbalanced Bridge

1. Purposes

(1) To understand the measuring principle of unbalanced bridge;

(2) To understand the concept of temperature coefficient of metal resistance;

(3) To practice two commonly used data processing methods (least squares and graphing methods);

(4) To understand the temperature dependence of semiconductor resistance.

2. Principle

(1) Principle of unbalanced bridge.

The measurement using the Wheatstone bridge have many advantages, such as high sensitivity, broad measurement range, high accuracy, simple circuit and easy to perform temperature compensation. The unbalanced bridge is widely used to measure the slowly varying physical quantities related with resistance, such as temperature, pressure and strain. The bridge can be classified into two kinds: DC Bridge and AC Bridge according to the source property.

A Wheatstone bridge is shown in Fig. 2.11.1, where R_1 is a measured resistor the value of which is related to temperature. U_0 is the voltage of half bridge DAC. So, the current through resistors R_1 and R_2 is

$$I_1 = \frac{U_0}{R_1 + R_2} \tag{2.11.1}$$

and the electric potential drop for R_2 is

$$U_{AC} = \frac{R_2}{R_1 + R_2} U_0 \tag{2.11.2}$$

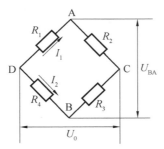

Fig. 2.11.1 Unbalanced bridge

In the save way, the electric potential drop for R_3 is

$$U_{BC} = \frac{R_3}{R_3 + R_4} U_0 \qquad (2.11.3)$$

The output voltage between the points A and B can be obtained from the difference of U_{AC} and U_{BC} as

$$U_{BA} = U_{BC} - U_{AC} = \frac{R_3}{R_3 + R_4} U_0 - \frac{R_2}{R_1 + R_2} U_0 \qquad (2.11.4)$$

When the bridge is in balance, namely $R_1 R_3 = R_2 R_4$ or $R_1/R_2 = R_4/R_3$, the output voltage of the bridge is zero, $U_{BA} = 0$ V.

Varying the temperature to change the resistance of R_1 and keeping the other resistors unchanged the bridge will be unbalanced and give a nonzero voltage between A and B. If $\Delta R_1 \ll R_1$, the output voltage caused by ΔR_1 can be written as

$$\Delta V = \frac{R_2 \Delta R_1}{(R_1 + R_2)^2} U_0 = \frac{R_2 R_1}{(R_1 + R_2)^2} \frac{\Delta R_1}{R_1} U_0 \qquad (2.11.5)$$

ΔV is proportional with ΔR_1.

When $R_1 = R_2 = R$, $R_3 = R_4 = R'$ (the bridge is called horizontal bridge) the expression can be simplified as

$$\Delta V = \frac{U_0}{4} \frac{\Delta R}{R} \qquad (2.11.6)$$

When $R_1 = R_4 = R$, $R_3 = R_2 = R'$ (the bridge is called vertical bridge) we have

$$\Delta V = \frac{R'R}{(R + R')^2} \frac{\Delta R}{R} U_0 \qquad (2.11.7)$$

When $R_1 = R_4 = R_3 = R_2 = R$ (the bridge is called equal arms bridge), the expression is the same as the horizontal bridge.

(2) Temperature dependence of resistance.

For a resistor made of metal or alloy, its resistance usually increases with the rising temperature. There is a linear relation between the resistance and temperature

$$R = R_0 [1 + \alpha(t - t_0)] \qquad (2.11.8)$$

where R and R_0 denote the resistance at the temperature t and t_0; α is the temperature coefficient of resistance (TCR), which has the units of inverse temperature, $1/°C$, or $°C^{-1}$. In fact, α is constant within a specific temperature range. When $t_0 = 0$ °C, there is

$$R = R_0(1 + \alpha t) \qquad (2.11.9)$$

Generally speaking, in 20 ~ 100 °C region, $\Delta R_1 \ll R_1$. Then we have

$$\Delta R = \alpha \cdot \Delta t \cdot R_0 \text{ or } \alpha = \frac{\Delta R}{R_0 \Delta t} \qquad (2.11.10)$$

From Eq.(2.11.6) and Eq.(2.11.10), it's easy to deduce the final relation for measuring the TCR:

$$\alpha = \frac{4}{U_0} \frac{\Delta V}{\Delta t} \qquad (2.11.11)$$

For a resistor made of semiconductor, its resistance usually decreases with the rising

temperature. There is an exponential relation between the resistance and temperature:
$$R = R_0 e^{E/KT} \quad (2.11.12)$$
where E and R_0 are constants; K is the Boltzmann constant; T is the temperature in Kelvin.

For the alloy resistor, the temperature effects on the resistance are very slight and the resistance variations can't be measured by multimeter, so we have to use the unbalanced bridge. But for semiconductor resistor, temperature has significant effect on the resistance, so we can directly measure the resistance using a multimeter.

3. Instruments

Two resistors (R_2 and R_3), one resistance box (R_4), power supply, digital voltage meter (DVM), multimeter, temperature controller with the alloy and semiconductor resistors in it, wires.

4. The contents of experiment

(1) Build a Wheatstone bridge according to Fig. 2.11.2.

Use alloy resistor as R_1, the resistance box as R_4, R_2 = 200 Ω, R_3 = 2 000 Ω.

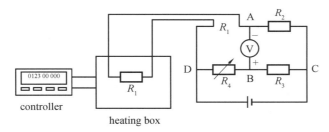

Fig. 2.11.2　　Experiment setup for measuring TCR using unbalanced bridge

(2) Measure the voltage of the power supply U_0 using a DVM. U_0 should be about 5 V.

(3) Set the heating box to 40 ℃, adjust the resistance box R_4 until the DVM reads zero, the bridge is balanced. Record the resistance of R_4, it is about 2 000 Ω. The full-scale of the DVM is 200 mV thereafter, and the voltage should have three decimal places.

(4) Rising the temperature of the heating box and read the voltage from the DVM for every 5 ℃ from 40 ℃ to 70 ℃.

(5) Measure the resistance of the semiconductor resistor using a multimeter for every 5 ℃ from 40 ℃ to 70 ℃.

5. Measurement and data processing

(1) Build a Wheatstone bridge, measure the voltage of the power supply U_0, set the heating box to 40 ℃, adjust the resistance box R_4 to balance the bridge and record R_4.

(2) Rising the temperature, and fill in Table 2.11.1 in the report. The first line in Table

2.11.1 is the measured voltage from the unbalanced bridge read from the DVM, and the second line is the resistances of the semiconductor resistor read from the multimeter.

(3) Calculate $\Delta V/\Delta t$ with two methods: graphing method and least squares method.

(4) Calculate the TCR of alloy resistor using Eq.(2.11.11).

(5) Plot the temperature dependence graph of the semiconductor resistance.

6. Discussion

(1) In the experiment, the TCR of alloy is positive or negative? What's that mean?

(2) Can you give other uses of unbalanced bridge?

Experimental Report

Measuring the Temperature Coefficient of Resistance Using Unbalanced Bridge

Name: Student ID No.:   Score:

Date: Partner: Teacher's signature:

1. Purpose

2. Measurement and data recording (Table 2.11.1)

Table 2.11.1 Measure the voltage and resistance

Temperature/℃	40.0	45.0	50.0	55.0	60.0	65.0	70.0
Voltage/mV							
Resistance/Ω							

Voltage of bridge, $U_0 =$

Resistance R_4 at 40 ℃, $R_4 =$

3. Calculate and plot

(1) The TCR of the alloy resistor.

① Calculate the $\Delta V/\Delta t$ by using of the least squares method: $k = \dfrac{\overline{x \cdot y} - \bar{x} \cdot \bar{y}}{\overline{x^2} - \bar{x}^2}$.

② Calculate the $\Delta V/\Delta t$ by using of the graphing method.

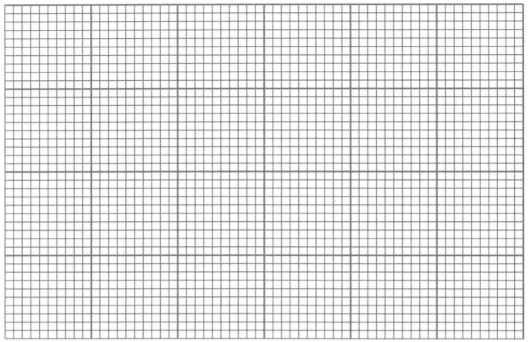

③ Calculate the TCR of the alloy resistor by using of $\Delta V/\Delta t$ and Eq.(2.11.11).

(2) Plot the temperature dependence of the semiconductor resistance.

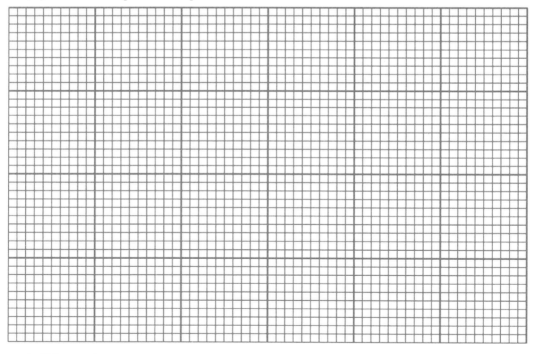

4. Discussion

2.12 Potentiometer and Its Application

The potentiometer (or potentiometer) is a high precision measuring instrument based on compensation principle, whose advantage is exhibited in some aspects, e.g. there is no influence on the measured system when it is in compensated status, which means, no current flows into or out of the measured system. With its higher accuracy and precision, potentiometer is often used to directly measure the electromotive force (or electromotance, electrodynamic force, electromotive potential) and voltage, intermediately measure the current, resistance and other electrical quantities. It is also often used to verify a variety of accurate meters. In scientific research and engineering, it finds wide application in measurement of nonelectrical quantities such as temperature, translation in cooperation with various sensor or transducer. In recent years, with the development of integrated circuit and computer technology, the potentiometer is gradually replaced by the higher precision digital meters, however, the investigation of physical idea on compensation principle and its application in measurement technique still have important significance.

1. Purpose

(1) To master the principle to measure the electromotive force by using compensation principle;

(2) To master the measuring technique using a potentiometer to study the output of a battery or primary cell.

2. Principle

(1) Compensation principle.

In a simplest circuit composed of two batteries and a galvanometer G, as shown in Fig. 2.12.1, when an electromotive force E_x is equal to another electromotive force E we say that they are in balance or in compensation state. If one adjusts the output of a known electrical source to balance an unknown electromotive force to make the galvanometer G indicate zero, then the unknown electromotive force can be quantitatively determined by the electromotive force of the known source. In the compensation state, there is no mutual effect of E and E_x on each other and no current flows from E_x into E, and vice versa. In other words, the circuit is like the one in the off state as if there is no connection between E_x and E.

In practice, one often utilizes the simple circuit to construct a variable source. As shown in Fig. 2.12.2, the part below the dashed line can be viewed as an adjustable source, which is composed of a main circuit with voltage divider and its output is the potential drop between the two changeable tapped resistances. In a real case for convenience one uses two resistances R_1, R_2: R_1 is coarse (rough) adjustment resistor with a gear position of resistance output, R_2 is fine adjustment resistor with output of continuous change of resistance. When the

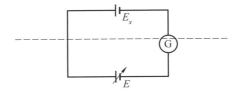

Fig. 2.12.1　A simplest circuit for compensation, two electromotive forces seem independent or isolated

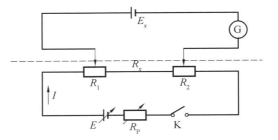

Fig. 2.12.2　A simple circuit to construct an electrical source with variable output from the tap resistances

galvanometer indicates zero, the potential drop just compensates the measured electromotive force. We have

$$E_x = IR_x \qquad (2.12.1)$$

where R_x is total output between the tap resistances used to balance E_x. If R_1, R_2 are readable with higher reading accuracy and I is adjusted to a precise value (standard current), E_x can be accurately determined.

(2) Operational principle of potentiometer.

Fig. 2.12.3 gives the schematic graph of a real potentiometer, in which the part enclosed by the dashed lines is consolidated to a black box and in the experiment the students should connect it with conducting wires to some necessary elements such as the electrical source E_x to be measured, standard battery E_N, galvanometer G, and variable electrical power supply E with a key switch K_1 and the adjustable resistor R_P.

The operation procedure is divided into two steps.

In the first step, the mode switch K is pushed to the right side, a source that outputs from a precise resistance with a fixed value (standard or normalization resistor) R_N starts to balance a standard battery or cell E_N. The aim of the step is to standardize or normalize the current I in the main circuit, which is constructed with a variable source E, a switch K_1, an adjustable resistance R_P, a normalization resistance R_N and a series of resistances R_1 and R_2. One should regulate the relation of all elements in the main circuit in order to make the current I reaching a standard current. In the real practice, normalization resistor R_N just takes a value that matches the value of E_N, asindicated in Fig. 2.12.3, to let E_N/R_N become some

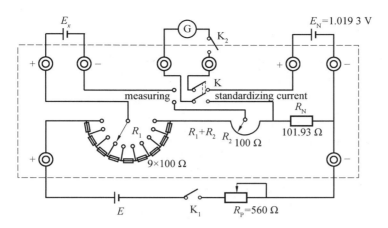

Fig. 2.12.3 Electrical scheme of a real potentiometer used in the students' experiments

value to facilitate the calculation of E_x, say one hundredth or one thousandth of an ampere (10 mA or 1 mA), one may notice that the significant figures are the same for E_N and R_N.

In the second step, the mode switch K is pushed to the left side, a variable source that outputs from tap resistors R_1 and R_2 is activated to measure the unknown electromotive force E_x. After roughly adjusting the tap position of R_1, the galvanometer G may approach the zero position already. Next carefully rotate the output of fine resistor R_2 until the G completely indicates zero position. Then the measured electromotive force E_x can be determined by a product of tap resistor R_x used and standard current I. In fact, since I is fixed in the first step already, the position regulation of R_1 and R_2 cannot bring any change in the main circuit. In the balance condition, it may be viewed as independent from outside environment, so we have

$$E_x = IR_x = \frac{E_N}{R_N}R_x \qquad (2.12.2)$$

Note that the mode key has off state, students should always let the key keep in that state when they don't want to perform any operation or they find that the instrument could not reach the balance anyhow. In the latter case, they should analyze the reason.

3. Experiment contents

(1) Repetitively (5 times) measure the electromotive force E_x of a D size battery paying attention to the reading from R_2, and calculate the average value.

(2) Study the construction and output characteristics of primary battery (galvanic cell) using table vinegar and some pieces of metal of different kinds (copper/iron nails or slices).

4. Discussion

(1) Please give the design to determine the internal resistance of a battery.

(2) Please list the causes of imbalance (unbalance) of a potentiometer.

5. Cautions

(1) In the experiment, it is not allowed that the standard battery is abruptly moved or put aside obliquely, even upside down.

(2) It is strictly forbidden to directly put the standard battery in circuit as an output source.

(3) The short circuit must be avoided.

Part 2　Topics of Experiments

Experimental Report

Potentiometer and Its Application

Name:　　　　　　　Student ID No.:　　　　　Score:

Date:　　　　　　　Partner:　　　　　　　　Teacher's signature:

1. Purpose

2. The experimental data and treatment

(1) Output measurement of a battery of D size (Table 2.12.1).

Table 2.12.1　**Measurement of output of a battery**

Measurement times	R_1/Ω	R_2/Ω	$(R_1+R_2)/\Omega$	I/A	E_x/V
1					
2					
3					
Mean value $\bar{E}_x/V$					

The calculation example for getting the significant figures of E_x and $\bar{E}_x$.

(2) Study the output of a primary cell.

①Table 2.12.2, the effect of volume of solution on output (make table by yourself).

113

②Table 2.12.3, the effect of metal area submersed in solution on output (make table by yourself).

③Table 2.12.4, the effect of kind of solution on output (such as cola, salt solution, make table by yourself).

3. Conclusion on the output characteristics of the primary cell

4. Discussion

2.13 Temperature Characteristic of PN Junction

1. Purpose

(1) To know the influence of temperature on the volt-ampere characteristics of PN junction;

(2) To learn the principle of temperature measurement of PN junction temperature sensor.

2. Principle

(1) The principle of temperature measurement of PN junction temperature sensor.

A PN junction is a boundary or interface between two types of semiconductor materials, P-type and N-type, inside a single crystal of semiconductor. The "P" (positive) side contains an excess of holes, while the "N" (negative) side contains an excess of electrons in the outer shells of the electrically neutral atoms there. This allows electrical current to pass through the junction only in one direction.

In forward bias, the P-type is connected with the positive terminal and the N-type is connected with the negative terminal and the holes in the P-type region and the electrons in the N-type region are pushed toward the junction and start to neutralize the depletion zone, reducing its width. The positive potential applied to the P-type material repels the holes, while the negative potential applied to the N-type material repels the electrons. The change in potential between the P side and the N side decreases or switches sign. With increasing forward-bias voltage, the depletion zone eventually becomes thin enough that the zone's electric field cannot counteract charge carrier motion across the PN junction, which as a consequence reduces electrical resistance. Electrons that cross the PN junction into the P-type material (or holes that cross into the N-type material) diffuse into the nearby neutral region. The amount of minority diffusion in the near-neutral zones determines the amount of current that can flow through the diode.

Only majority carriers (electronsin N-type material or holes in P-type) can flow through a semiconductor for a macroscopic length. With this in mind, consider the flow of electrons across the junction. The forward bias causes a force on the electrons pushing them from the N side toward the P side. With forward bias, the depletion region is narrow enough that electrons can cross the junction and inject into the P-type material. However, they do not continue to flow through the P-type material indefinitely, because it is energetically favorable for them to recombine with holes. The average length an electron travels through the P-type material before recombining is called the diffusion length, and it is typically on the order of micrometers.

Although the electrons penetrate only a short distance into the P-type material, the electric current continues uninterrupted, because holes (the majority carriers) begin to flow in the opposite direction. The total current (the sum of the electron and hole currents) is constant in space, because any variation would cause charge buildup over time (this is Kirchhoff's current law). The flow of holes from the P-type region into the N-type region is exactly analogous to the flow of electrons from N to P (electrons and holes swap roles and the signs of all currents and voltages are reversed).

Therefore, the macroscopic picture of the current flow through the diode involves electrons flowing through the N-type region toward the junction, holes flowing through the P-type region in the opposite direction toward the junction, and the two species of carriers constantly recombining in the vicinity of the junction. The electrons and holes travel in opposite directions, but they also have opposite charges, so the overall current is in the same direction on both sides of the diode, as required.

The forward current I_F and forward voltage of a PN junction satisfy the following relationship:

$$I_F = I_n e^{\frac{qV_F}{kT}} \quad (2.13.1)$$

where q is the amount of electron charge (positive value); T is thermodynamic temperature; k is Boltzmann constant; I_n is the reverse saturation current, which is related to the forbidden band width and temperature of PN junction.

$$I_n = CT^\gamma e^{-\frac{qV_{g(0)}}{kT}} \quad (2.13.2)$$

where C is a constant related to the junction zone of boundary or interface between two types and doping concentration of PN junction; γ is a constant related to crystal of semiconductor; $V_{g(0)}$ represents the potential difference between the bottom of conduction band and top of valence band. $V_{g(0)}$ is a constant for a given material. Substituting Eq. (2.13.2) into Eq. (2.13.1) and taking the logarithm of both sides, one can get

$$V_F = V_{g(0)} - \frac{kT}{q}\ln\frac{C}{I_F} - \frac{kT}{q}\ln T^\gamma \quad (2.13.3)$$

Eq. (2.13.3) goes into the relation among forward voltage, forward current and temperature, which is a basic equation of the temperature sensor of PN junction. On right side of equal sign, the first term is a constant, the second term is a linear term for temperature, and the third term is a non-linear term for temperature. Some experiments show that the nonlinear term can be ignored when the temperature range is small. For example, for the conventional silicon-based PN junction, if temperature change from −50 ℃ to 150 ℃, the influence of nonlinear term is less than 0.5% in contrast with linear term. Thus, Eq. (2.13.3) becomes

$$V_F = V_{g(0)} - \left(\frac{k}{q}\ln\frac{C}{I_F}\right)T \tag{2.13.4}$$

If forward current I_F is kept constant, the temperature is directly proportional to forward voltage of PN junction, which means we can measure temperature by measuring forward voltage of PN junction. This is the basis principle of PN junction temperature sensor.

T in the above equations denotes the thermodynamic temperature and its unit is Kelvin. In order to conduce to actual measurement, we convert it into Celsius. The relation between them is $T = 273.15 + t$, with t representing Celsius temperature. Eq.(2.13.4) is changed to

$$V_{F(t)} = V_{g(0)} - \left(\frac{k}{q}\ln\frac{C}{I_F}\right)(273.15 + t) = V_{F(0)} - \left(\frac{k}{q}\ln\frac{C}{I_F}\right)t \tag{2.13.5}$$

where $V_{F(t)}$ represents the forward voltage of PN junction at t (℃) and $V_{F(0)}$ represents the forward voltage of PN junction at 0 ℃. It needs to be pointed out that the current passing through the PN junction should be kept constant when using the PN junction to measure temperature. One defines the sensitivity of PN junction sensor as

$$S = \frac{k}{q}\ln\frac{C}{I_F} \tag{2.13.6}$$

Therefore Eq.(2.13.5) is rewritten as

$$V_{F(t)} = V_{F(0)} - St \tag{2.13.7}$$

If we let $\Delta V = V_{F(t)} - V_{F(0)}$, Eq.(2.13.7) also can be written as

$$t = \frac{-\Delta V}{S} \tag{2.13.8}$$

(2) Forbidden band width of PN junction material.

For a particular PN junction material, forbidden band width is important, which is defined as

$$E_{g(0)} = qV_{g(0)} \tag{2.13.9}$$

where $E_{g(0)}$ represents the forbidden band width; $V_{g(0)}$ is the difference of potential between the bottom of conduction band and top of valence band at 0 ℃. Since the measurement in this experiment start from room temperature, the forward voltage of PN junction measured at room temperature can be used to calculate $E_{g(0)}$. If t_0 stands for the room temperature, according to equation Eq.(2.13.4), we get $V_{g(0)} = V_{F(t_0)} + S(273.15 + t_0)$ and

$$E_{g(0)} = qV_{g(0)} = q[V_{F(t_0)} + S(273.15 + t_0)] \tag{2.13.10}$$

3. The experiment content

(1) Measure the forward volt-ampere characteristic curve of PN junction.

In Eq.(2.13.1), for a fixed temperature, the forward current I_F and the voltage V_F of PN junction take on exponential relationship, as shown in Fig. 2.13.1.

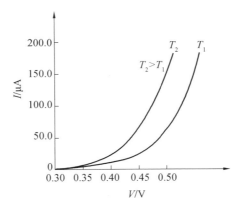

Fig. 2.13.1 Volt-ampere characteristics of PN junction

In order to present this relation, let V_F change from 0.30 V to 0.50 V with interval being 0.02 V and record I_F at room temperature and 40 ℃ respectively. (Fill in the blanks in Table 2.13.1 in the experimental report.) Then plot the V - A curves of I_F - V_F in graphic paper. (Draw the two curves in the same coordinate and intuitively understand the influence of temperature on volt-ampere characteristics.)

(2) Measure the change of forward voltage with temperature keeping current constant.

①Keep forward current being 50 μA and measure the forward voltage every 5 ℃ from 40 ℃ to 85 ℃. (Fill in the blanks in Table 2.13.3 in the report.) Don't forget to record V_F at room temperature because we will use it to calculate forbidden band width below. (Fill in the blanks in Table 2.13.3 in the report.) Plot the V_F - t curve in graphic paper. Please note that I_F will change slightly when temperature increases, so adjust it back to 50 μA when you record V_F.

②Take two points on the V_F - t curve and calculate the sensitivity by

$$S = -\frac{V_2 - V_1}{t_2 - t_1}$$

(3) Determine the forbidden band width $E_{g(0)}$.

①Calculate the electric potential difference $V_{g(0)}$ at 0 ℃ by

$$V_{g(0)} = V_{F(t_0)} + S(273.15 + t_0)$$

②Calculate the forbidden band width $E_{g(0)}$ with $V_{g(0)}$ by

$$E_{g(0)} = qV_{g(0)}$$

4. How to use heater

On the display window, there are two rows of figures with different color. The red upper figures are the temperature of the furnace chamber of the heater, namely the temperature of PN junction, which is measured by Pt100 temperature sensor inside the chamber. The green lower figures show the preset temperature of the heater to be heated, which is manually set by the buttons below.

When intelligent radiant-type heater turned on, the preset temperature (green figures) should be less then room temperature to ensure that the heater is not in a state of heating. At this time, the measured temperature (red figures) is room temperature. Thus, while you fulfill this experiment, make sure that the preset temperature (green figures) is under room temperature. For example, it can be set as −10 ℃.

5. Discussion

(1) When we measure the change of forward voltage with temperature, why should we take I_F a small value and keep it constant?

(2) Search related materials to elaborate what is PN junction.

Experimental Report

Temperature Characteristic of PN Junction

Name:　　　　　　　Student ID No.:　　　　　　Score:

Date:　　　　　　　Partner:　　　　　　　　　Teacher's signature:

1. Purpose

2. The experimental data and processing

(1) Measure the forward volt-ampere characteristic curve of PN junction (Table 2.13.1).

Table 2.13.1　Forward current for different voltage

Room temperature	V_F/V	0.30	0.32	0.34	0.36	0.38	0.40	0.42	0.44	0.46	0.48	0.50
	I_F/mA											
40 ℃	V_F/V	0.30	0.32	0.34	0.36	0.38	0.40	0.42	0.44	0.46	0.48	0.50
	I_F/mA											

Plot the V − A curves (namely I_F − V_F curves) at room temperature and 40 ℃ respectively in graphic paper below according to Table 2.13.1.

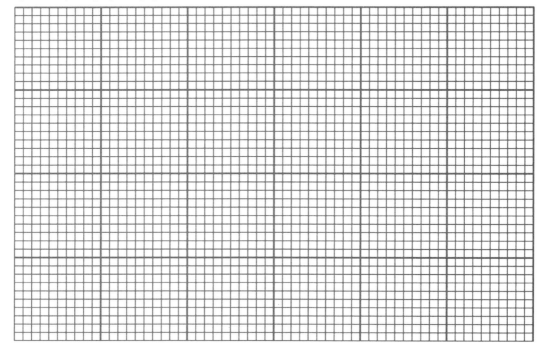

Part 2 Topics of Experiments

(2) Measure the change of forward voltage with temperature keeping current 50 μA (Table 2.13.2 – 2.13.3).

Table 2.13.2 Forward voltage at room temperature

Room temperature	
$V_{F(t_0)}/V$	

Table 2.13.3 Forward voltage for different temperature

t ℃	40	45	50	55	60	65	70	75	80	85
V_F/V										

Plot the $V_F - t$ curve in graphic paper according to Table 2.13.3.

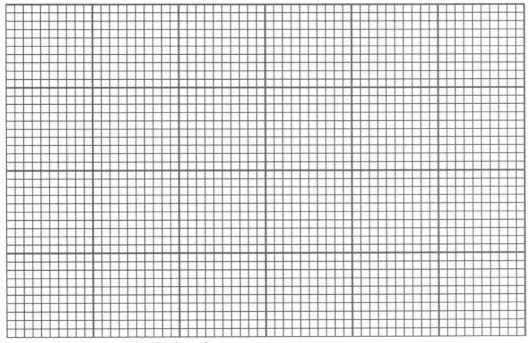

Take two points and calculate the sensitivity.

$$S = -\frac{V_2 - V_1}{t_2 - t_1} =$$

(3) Calculate the forbidden band width $E_{g(0)}$ and relative error.

① Calculate the electric potential difference $V_{g(0)}$ according Table 2.13.2 and sensitivity S.

$V_{g(0)} = V_{F(t_0)} + S(273.15 + t_0) =$

$S =$

②Calculate the forbidden band width $E_{g(0)}$ and relative error with $V_{g(0)}$.

$E_{g(0)} = qV_{g(0)} =$

3. Discussion

2.14 Oscilloscope and Its Applications

Any change of a physical quantity (such as mechanical, thermal, sonic, optical electrical quantities) varying with time can be converted into electronic signal. Some meters such as digital multi-meter (DMM) are usually used to determine the voltage and current of a signal but failed in giving the information on how they vary in the time domain and exhibit the parameters (such as amplitude, period, and frequency) describing the variation characteristics, even less information in frequency domain such as frequency components or the spectrum of a signal.

Oscilloscope is a wide used scientific instrument that can perform multi-function of some meters especially in time or frequency domain. A periodical signal even pulsed signal can be displayed, stored, even analyzed on the oscilloscope.

There are mainly two kinds of oscilloscopes: analogue ones and digital ones. The former are used to demonstrate the periodic signals while the latter to show the transient or non-periodical signals, such as pulses, and store the signals except the periodic signals. Note that in the recent years by making use of the computing and sensor technology, the new electronic technique — the virtual instrument technology, including the virtual oscilloscope or network oscilloscope, is developed. In this experiment, we mainly introduce the analogue oscilloscope.

The electronic control and the display, as the two parts, such as CRT (cathode ray tube), sometimes LCD (liquid crystal screen) may construct the main structure of an oscilloscope.

1. Purpose

(1) To be familiar with two common lab instruments — the function generator and oscilloscope which are used to produce, display and measure the temporally varying electrical signals;

(2) To practice the observation, measurement of any periodic signal;

(3) To reveal the frequency relation of an interval between musical tones using oscilloscope (selective).

2. Principle

(1) The basic structure and operation fundamentals of an analogue oscilloscope.

①Basic structure of analogue oscilloscope. The basic structure of an analogue oscilloscope generally consists of CRT and the electronic control part, as shown in Fig. 2.14.1.

Any signal from the input port (Y-input, X-input, or the synchronized input) may first experience amplification or attenuation (amp./atten.), and through some switches (S_1, S_2) is finally applied to (loaded onto) the paired-plates in horizontal or vertical directions inside the CRT to control the two-dimensional movement of electron beam emitted from the electron gun in the end of CRT (Fig. 2.14.2).

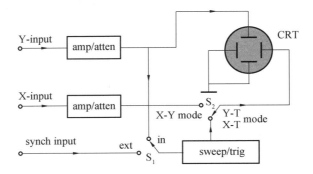

Fig. 2.14.1 The basic structure of an analogue oscilloscope

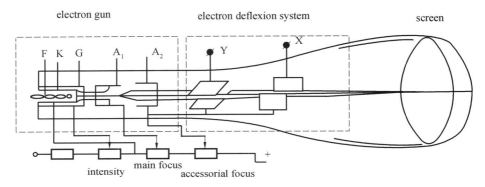

Fig. 2.14.2 The basic structure of a CRT

Usually there are two operation modes for an analogue oscilloscope: Y-T (X-T) mode or X-Y mode, in the former the signal with sawtooth waveform from the sweep/trigger unit is applied to the horizontal plates which cooperatively demonstrate the waveform of the signal applied to the vertical plates. While in the latter, the signal from X-input port (Channel X, CH X or CH 1 on an oscilloscope) or the Y-input port (Channel Y, CH Y or CH 2) may directly control the resultant motion of the electron beam sometimes. If there is a stable integer ratio in the frequency between vertical signal and horizontal signal the motion may be indicated in an enclosed curve.

②The structure of a CRT.

The CRT's structure is indicated in Fig. 2.14.2.

The electrons are emitted from filamentary cathode F and accelerated by the gratings K and G, experiencing the focus (by main focus A_1 and accessorial focus A_2) finally form a fine beam passing through zone of the vertical and horizontal plates and bombard the phosphor on the screen. By using effect of persistence of vision, the electron trace is displayed. The two pair of plates form the deflection system of the electron beam, the applied voltages on which finally control the graph on the screen.

Any periodic signal applied to the vertical or horizontal deflection plates may result in a vibration of electron in a certain direction. When the electron beam reaches the screen it leaves a trace, a bright straight line if the vibrational frequency become higher enough to follow (this is the effect of persistence of vision), as shown in Fig. 2.14.3.

Fig. 2.14.3 A faster vertical periodic signal forms a straight line on the screen

To demostrate how a signal $u = u(t)$ varies with time t one should "open" it in space so that the horizotal axis act as the temporal axis. Thus a signal increasing with time t (proportional to t) $u(t) = kt$ is needed to be applied to the horizontal deflection plates in order to generate a composite (resultant) motion for the electron beam. In the real case, it is impossible to get a signal which increases linealy with t to infinity. We often use a so-called sawtooth waveform to pull signal open in horizotal direction. The cooperation of sawtooth waveform with any signal can demostrate the signal's variation with time. Fig. 2.14.4 shows how the sawtooth waveform opens a sine signal with the same period on the screen. In fact, after the first period, the traces are drawn repeatedly starting from the same position, which directly leads to steady display of signal waveform on the screen, so we often call the sawtooth waveform the "sweep signal" or "scanning signal". When the period of signal T_y strictly martches that of sawtooth waveform T_x, namely $T_x = T_y$ (or $f_x = f_y$), one may see a whole signal's waveform of a period on the screen. In fact, at the end of a period of the sawtooth waveform, the scanning is repeated again from the starting point. If $T_x = nT_y (n = 1, 2, 3, ...)$ or $f_x = f_y/n$ then a complete waveform of signal with n periods can be depicted on the screen. Actually only if n is an integer the stability of demostration of the periodic signal is guaranteed. When there is no integel relation in periods or frequencies between the diaplayed signal and sawtooth waveform, the signal displayed on the screen will not be stable, sometimes, seems moveble or running.

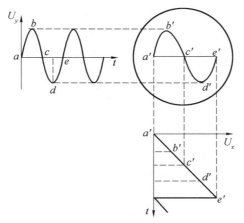

Fig. 2.14.4 The demonstration process of a signal with assistance of sawtooth waveform with the same period — the resultant motion of the electron beam on the screen

③The trigger (driving) sweep and synchronization. Actually the sawtooth signal may be generated inside the oscilloscope independent of the arbitrarily external signal to be displayed. How to control it to keep an integer ratio with the signal in the period or frequency to make the signal's display steady is the key point for different types of analogue oscilloscopes. Most oscilloscopes employ the method of "trigger sweep" or "forced sweep" to keep the stability of signal's display. In fact, the sampled signal from a given signal to be displayed triggers or drives the generation of the sweeping sawtooth signal. The sampling is completed by adjusting the position of triggering level in the trigger circuit (Fig. 2.14.5). According to the positions of the pulses formed by sampling, the starting and the ending points of the sawtooth waveform are determined. By using the phase shifting technique, a sweeping signal fitting the displayed signal in period is generated and applied onto the horizontal plates.

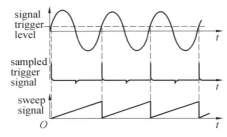

Fig. 2.14.5   The generation of trigger signal sampled from a given signal and the sweep sawtooth waveform

Some sort of oscilloscope, which may operate in so called "synchronization" mode, may force the displayed signal to match an external trigger signal in order to keep an integer ratio in period or frequency. It is necessary for synchronizing signal (external trigger signal) to reach a certain level.

(2) Signal observation and determination of its parameters.

①Display and observation of an arbitrary external signal. By using BNC (Bayonet Nut Connector) co-axial cables or Banana plug cables with BNC adapter the arbitrary external or internal signal such as from the function generator, the converter or sensor of any non-electric signal, can be applied to the input port CH X or CH Y. By correctly choosing the properties of the signal, display mode, trigger source and trigger mode and regulating the sensitivity (the Volt/Div knob to control attenuation and amplification of the signal), sweep period and trigger level, the signal may be moderately and stably displayed in the screen.

②Record of characteristic parameters for a periodic signal.

a. The amplitude. In order to measure the amplitude of a signal one needs first to adjust the sensitivity in CH Y or CH X and the signal position so that the signal is displayed in the frame of scales on the screen. Then one should record the number of grids occupied by the signal in the scales, thus $V_{p\text{-}p}$ the reading difference between the peak and valley in vertical direction can be calculated by considering the number multiplied by the value of sensitivity.

Part 2 Topics of Experiments

Note that the number of occupied grids should be evaluated to the fractional part of a grid. For instance, a signal with occupied 5.3 grids displayed in a sensitivity of 2 V/div has the amplitude in height between the peak and valley $V_{p-p} = 5.3 \times 2 = 10.6$ (V).

b. The period. Similarly to the above procedure, one first needs to read out the width of a signal's period in horizontal direction (the grid number occupied by a period) then consider sweep sensitivity (the reading can be obtained from the sweep time regulation knob, Times/div, or from a compact LC display) and the period can be written out. For example, if one period of signal occupies 5.0 grids in horizontal direction and the sweep sensitivity is 25 ms/div, then the period of the signal $T = 5.0 \times 0.25 = 1.25$ (ms).

c. The frequency. After getting the period T the frequency f of a signal can be easily calculated by

$$f = 1/T \qquad (2.14.1)$$

③Resultant motion of two signals. Lissajous figures are often met in some enclosed graphics, which are formed by two simple harmonic vibrations (signals) in the two perpendicular directions, say in x and y axis. The vibration in x axis may be expressed as $u_x = A_x\cos(\omega_x t + \varphi_x)$ while in y axis $u_y = A_y\cos(\omega_y t + \varphi_y)$, where $A_{x,y}$, $\omega_{x,y}$ and $\varphi_{x,y}$ denote the amplitude, frequency and initial phase in the corresponding directions. When the two vibrations have an integer frequency ratio, namely $\omega_y/\omega_x = n_x/n_y$ (n_x, n_y is an integer) and a fixed phase difference, namely $\Delta\varphi = \varphi_y - \varphi_x =$ constant, the resultant motion forms the enclosed trace — Lissajous figure. Such kinds of figures were first studied by a French scientist Lissajous so the figures are named after him.

In fact, Lissajous figures provide us a method to determine frequency or period of an unknown signal from the known signal if they compose the Lissajous figure. Fig. 2.14.6 shows the Lissajous figures for some ratio $\omega_y/\omega_x = n_x/n_y$ with a fixed phase difference $\Delta\varphi = \varphi_y - \varphi_x$.

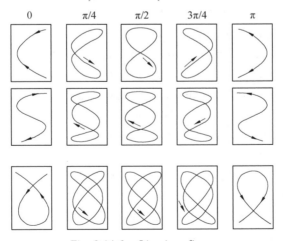

Fig. 2.14.6 Lissajous figures

(The upper, middle and lower row correspond to $\omega_y/\omega_x = $ 1/2, 1/3, 2/3 each figure corresponds to the phase difference $\Delta\varphi = \varphi_y - \varphi_x$ of 0, $\pi/4$, $\pi/2$, $3\pi/4$, π)

Actually some graphics are not enclosed but still belong to Lissajous figures called as degenerate ones (see the case with phase difference of $0, \pi$). There is a method which can be used to determine the ratio of frequencies by the ratio of maximum numbers of tangential points n_y/n_x in the two directions, namely $\omega_y/\omega_x = n_x/n_y$. It should be paid attention that the more tangential points obtained in a certain direction mean the lower frequency of signal loaded to this direction. We may notice that the method to count the numbers of tangential points is not suitable for the degenerate cases. The method of counting the maximum numbers of secant points in the two perpendicular directions may be the universal one (you may have a try to determine the ratio of frequencies for any case, especially degenerate case)!

3. Apparatuses

Oscilloscope (Fig. 2.14.7 and Fig. 2.14.8 in Appendix).
Function generator (Fig. 2.14.9 and Fig. 2.14.10 in Appendix).
BNC co-axial cables.
"T" adapter.
Banana plug cable with BNC adapter.
Black box with multiple output.
Microphone.
Electronic amplifier.
Musical instrument (electronic synthesizer, flute, guitar), test tubes.

4. Contents of experiment

(1) Please try to display in any channel (CH1 or CH2) a 500 Hz signal from function generator, and make it steady and determine its amplitude, period and frequency using the method introduced in the context. Change the sensitivity for amplitude and time (Volt/Div and Time/Div) to see the influence on display and final calculation results.

(2) Please try to display two signals from two function generators with the same frequency of 500 Hz in DUAL mode and in ADD mode (cooperation between two students).

(3) Please in X-Y mode try to composite Lissajous figures with ratio of frequencies 1:2, 1:3, 3:2, 3:4, 4:5, 8:9 proving the relation with the ratio of numbers of tangential and secant points. In DUAL mode try to display simultaneously the two signals (cooperation between two students).

(4) Study the output from the different port of a black box.

(5) Study the musical sounds (selective).

A musical tone (or sound) is the amazing thing that differs from any noise. As a mechanical vibrational signal, it has the periodicity or quasi-periodicity. You may know that some combinations of musical tones possess the harmony or consonance, but why? As Pythagoras said, there are relation of integer ratio of pitches between a pair of tones in a musical scale. Have you ever expected this phenomenon? Measure the frequency of tones of

Part 2 Topics of Experiments

musical scales in one or two octaves. You can try to study their relations and you will learn more that you have never noticed or realized.

5. Appendix

(1) The front panel of the dual trace oscilloscope (GOS-620) (Fig. 2.14.7).

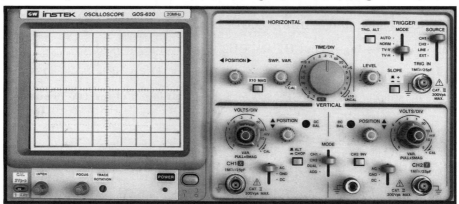

Fig. 2.14.7   The front panel of dual trace oscilloscope (GOS-620)

Main functions of the keys, knobs and display screen on panel are as follows.

①Below Screen.

$\boxed{\text{POWER}}$ button. Push the button to turn on the power, and the display (LED) is activated. Push again the button to turn off the power.

$\boxed{\text{INTEN}}$ knob is to control the intensity of trace and lightspot.

$\boxed{\text{FOCUS}}$ knob is to control the focusing of the trace or lightspot (electronic beam).

$\boxed{\text{TRACE ROTATION}}$ is to regulate the degree of parallelism of horizontal trace with the scale mark on screen.

② $\boxed{\text{VERTICAL}}$ is the deflection control.

$\boxed{\text{VOLTS/DIV}}$ gear knob for CH1 and CH2 is the gain/attenuation control (from 5 mV/Div to 5 V/Div).

In the middle, there is a small knob which can regulate the sensitivity of the input. In its $\boxed{\text{CAL}}$ position the sensitivity is not changed. When pulling out this knob (the status of × 5 MAG), the vertical sensitivity increases 5 times.

$\boxed{\text{AC-GND-DC}}$ switch is to choose the signal property.

$\boxed{\text{CH1(X)}}$ / $\boxed{\text{CH2(Y)}}$ BNC the terminal to link the external signal, in the default, CH1 corresponds to X deflection plates while CH2 to Y deflection plates.

$\boxed{\blacktriangle \text{POSITION} \blacktriangledown}$ knob is to control the trace position in vertical direction.

129

Vertical MODE switch is to select the display mode for single channel display (CH1 or CH2) or dual channel display (DUAL status, CH1 and CH2) and the summation display (CH1 + CH2, ADD status).

CH1&CH2 DC BAL is to regulate the DC balance position in vertical direction.

ALT/CHOP button in the DUAL MODE releasing this button CH1 and CH2 display alternatively while pressing it the CH1 and CH2 display in a chopped way.

CH2 INV button. When the button is pressed the CH2 signal is inversed. In the ADD mode, the summation of CH1 and CH2 suggests the subtraction of the two channels (CH1 − CH2).

③ HORIZONTAL is the deflection control.

TIME/DIV knob is to select the sweep time or control the frequency (period) of scanning sawtooth waveform. The extreme end X-Y set the X-Y mode in which the signals from CH1 and CH2 may composite their motion in 2-dimensional way on the screen (such Lissajous figures).

SWP.VAR. knob is control for the variable sweep. In its CAL position, there is no variation in sweep frequency.

◄POSITION► knob is to control the trace position in horizontal direction.

④TRIGGER Control.

SOURCE switch is to select the trigger from inside (CH1 or CH2) or from outside (line or external).

In the vertical display mode in the DUAL or ADD state, choosing CH1 the signal from CH1 acts as internal trigger source while choosing CH2 the signal from CH2 as the trigger source. Choosing LINE status the frequency of trigger signal becomes the frequency of the electric supplying line. Choosing EXT status the external signal from EXT TRIG. IN BNC input port acts as the trigger signal.

Trigger MODE switch.

On the AUTO state when there is no trigger signal whose frequency is lower than 25 Hz, the sweeping will generated automatically.

On the NORM state with no trigger signal the instrument is located in the ready state for sweeping. There is no trace on the screen; this function is used to observe the signal below 25 Hz.

The TV-V or TV-H state is used to measure the signal of a TV picture in vertical or horizontal direction.

Part 2   Topics of Experiments

SLOPE button is the select button of trigger slope button in released or pressed state corresponds to the trigger with positive slope or negative slope.

EXT. TRIG. IN BNC is the input terminal to linking to an external trigger source.

TRIG. ALT. button. When the vertical display mode is switched to DUAL or ADD state pressing the button the instrument automatically set the signal from CH1 and CH2 as the trigger in in the vertical display mode in the DUAL or ADD state, choosing CH1 the signal from CH1 acts as internal trigger source in the alternative way.

LEVEL is the knob to regulate the trigger standard position if rotating it to " + " direction the standard position will move upwards while rotating to " − " direction it will move downwards.

(2) The front panel of the dual trace oscilloscope (GOS-630FC) (Fig. 2.14.8).

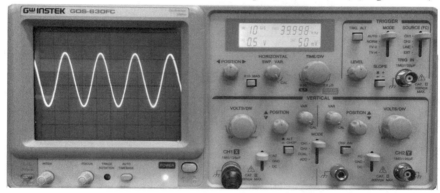

Fig. 2.14.8   The front panel of dual trace oscilloscope (GOS-630FC)

The oscilloscope GOS-630FC, in fact, is the updated version of GOS-620, which provides a piece of LCD to display the information on volt/div of each channel and information on time/div, and evaluated frequency of the displayed signal. In the control panel, the small knob for vary the magnification of signal of each channel, which is located in the middle of VOLTS/DIV for GOS-620, is individually designed and symmetrically arranged near the MODE switch.

(3) The front panel of the function generator (SFG-2010) (Fig. 2.14.9).

Fig. 2.14.9   The front panel of sychronic function generator (SFG-2010)

①Main functions of the keys, knobs and display LEDs on the panel.

POWER button.

Push the button to turn on the power, and the display is activated. Push again the button to turn off the power.

WAVE button is the main function key.

This key is to set main output waveform in the cycle of Sine, Triangle and Square. When the key is pressed, the related waveform LEDs will light up accordingly.

Entry keys.

Press the number key (1 2 ⋯ 9 0) and the dot key (.) to input value, then press Unit key (MHz KHz Hz%) to complete the value setting. SHIFT key is for the secondary function mode setting. When the key is pressed, the instrument will choose secondary function and the SHIFT LED will light up.

Unit keys.

MHz KHz Hz% keys are used to assign the unit and set the entered value of frequency (MHz, kHz, Hz) and duty cycle (%) in the normal mode.

②Modify keys.

◀ ▶ keys are used to change the digit of input value. Rotate the knob for increasing or decreasing that digit.

Waveform function LEDs indicate the figure of main output waveform and the current operation functions.

Other LEDs indicate the status of the wave, such as Unit indicate LED, Duty function LED, Attenuation indicate LED.

Main OUTPUT BNC.

This BNC connector is to output all main signals with 50 Ω output resistance.

TTL/CMOS OUTPUT BNC.

This BNC connector is to output TTL/CMOS compatible level signals.

AMPL knob (Output Amplitude Control) with Attenuation Operation.

Turnthe knob clockwise for MAX output and invert for Min output. Pull the knob out for an additional 20 dB output attenuation.

OFFSET is the control knob.

Pull out the knob to select any DC level of the waveform between ±5 V(into 50 Ω load), turn the knob clockwise to set a positive DC level waveform and invert for a negative DC level waveform.

TTL/CMOS is the selector knob.

When push SHIFT + 9 (for TTL) keys and push in the knob, the BNC terminal will output a TTL compatible waveform. Pull out the keys and rotate the knob can adjust the CMOS compatible output ($5\text{-}15V_{p\text{-}p}$) from the output of BNC.

(4) The front panel of the function generator (AFG-2125)(Fig. 2.14.10).

Fig. 2.14.10   The front panel of sychronic function generatorn (AFG-2125)

The function generator (AFG-2125) is the advanced version of signal source with arbitrary waveform design. The basic function is the same with the function generator (SFG-2010), only the distribution of the keys, knobs are different.

To choose the function of a signalin the cycle of Sine, Triangle, Square, Arbitrary waveform, press the button of FUNC.

To set the frequency before press number input key, press the button of FREQ.

To adjust amplitude of signal, press AMPL and rotate the biggest rotator.

To give the final output, press OUTPUT key, which will become bright.

6. Discussion

(1) How to make a waveform steady (not to move towards left or right side) on the screen?

(2) If you find a straight line on a screen in a certain direction and you are sure that the signal is accessed correctly. Please try to explain the phenomenon by listing the possible reasons.

(3) Putting the musical tones Do and Sol together they may sound more comfortable or consonant than Do and Ti, why? (selective)

Experimental Report

Oscilloscope and Its Applications

Name:　　　　　　　　Student ID No.:　　　　　　Score:

Date:　　　　　　　　　Partner:　　　　　　　　　Teacher's signature:

1. Purpose

2. The experimental procedures and data treatment

(1) Display a 500 Hz signal from function generator (any waveform), make it steady and determine its amplitude, period and recalculate frequency from the period (Table 2.14.1 ~ 2.14.2).

Table 2.14.1　For measuring basic parameters of a signal

Waveform	Amplitude/V	Period T/s	Frequency f/Hz
	grid reading × div/grid = 　× =	grid reading × div/grid = 　× =	$f = 1/T =$

Display the signal from a black box with several output ports.

Table 2.14.2　For measuring basic parameters of signal from black box

Waveform	Amplitude/V	Period T/s	Frequency f/Hz
	grid reading × div/grid = 　× =	grid reading × div/grid = 　× =	$f = 1/T =$
	grid reading × div/grid = 　× =	grid reading × div/grid = 　× =	$f = 1/T =$

Part 2   Topics of Experiments

Continued   Table   2.14.2

Waveform	Amplitude/V	Period T/s	Frequency f/Hz
	grid reading × div/grid = × =	grid reading × div/grid = × =	$f = 1/T =$
	grid reading × div/grid = × =	grid reading × div/grid = × =	$f = 1/T =$
	grid reading × div/grid = × =	grid reading × div/grid = × =	$f = 1/T =$

(2) Display two signals from two sources (the one is function generator another one is the black box) with the same frequency (How?) in DUAL mode and in ADD mode.

(3) Composite Lissajous figures with frequency ratio of 1∶2, 1∶3, 3∶2, 4∶3, 5∶4, 8∶9. (at least 3 cases) using the sine signals from function generator and the black box (select the port, which has the sine waveform output).

Try to draw the figures of each case (with fixed phase difference) and prove the relation of frequency and considering the real frequencies displayed in signal generators using method of tangent or secant line. Try to draw the figures and give the relations of frequency ratio.

(4) Study the musical sounds (selective).

①Observe the quasi-periodicity of a stable musical sound (simulating the any musical sound using electric keyboard, say a flute) and measure the frequency of each musical note in a musical scale of an octave by determining the periods(Table 2.14.3).

Table 2.14.3   For determing the frequency (in Hz) of musical notes in an octave

f_{Do}	f_{Re}	f_{Mi}	f_{Fa}	f_{Sol}	f_{La}	f_{Ti}	$f_{Do'}$

②In X-Y mode, use the signal from function generator with frequency of 523 Hz for Do in KEY C and the signal from the electric keyboard (select "flute") by playing Do, Re, Mi,

Fa, Sol, La, Ti, Do',... on it and try to composite the figures. Observe which combination will give clear and stable Lissajous figure (you may slightly regulate the frequency of a signal from the function generator (How?) to make the figure stable on screen) and try to find the frequency ratio, which is reflecting the integer ratio between the intervals of two musical notes. Tell the tutor what amazing thing you find and summarize it and write down something below!

3. Discussion

2.15 Measurement of Sound Velocity in Air

1. Purpose

(1) To be familiar with the travelling characteristics of sound wave via the measurement of velocity of ultrasound in air;

(2) To master the use of the signal generator and the oscilloscope completely;

(3) To practice the method of successive difference to process data.

2. Principle

Sound is a mechanical wave which may travel in the elastic medium. It belongs to the sort of longitudinal waves whose vibrational direction is coincident with its propagation direction. The sound in a frequency range from 20 Hz to 20 kHz is the audible wave, which can stimulate the reaction of human auditory system. While the sound higher than 20 kHz, which is specially termed as ultrasonic wave or ultrasound, cannot be heart by human ears. Compared to the audible sound the ultrasound possesses one of the most obvious characteristics — travelling directionality, namely the ultrasonic beam may travel directionally to a fixed direction. It has stronger penetrability, the attenuation of ultrasonic wave is much smaller than audible sound in liquids and solids. Reflection may occur when it encounters the interface composed of two media with different density, which forms the technique basic for detecting the defects or cracks, distance-ranging, echolocation, even sectionally analyzing the human body (Computed Tomography, CT).

The travelling speed of sound in air is

$$v = \sqrt{\frac{\gamma RT}{m}} \qquad (2.15.1)$$

where γ denotes the ratio of specific heat capacity, $\gamma = C_p/C_V$ with C_p, C_V being the heat capacity at constant pressure and the heat capacity at constant volume; R is universal gas constant; m is average mass of an air molecule; T is the thermodynamic temperature. The parameters γ, m are strongly affected by the air composition, especially the humidity (the contents of water vapor). At the sea level, at 0 ℃, under the standard pressure of atmosphere, the sound velocity in air v_0 = 331.45 m/s, so the sound velocity at t ℃ is

$$v = v_0 \sqrt{\frac{T}{273.15}} = 331.45 \sqrt{1 + \frac{t}{273.15}} \text{ (m/s)} \qquad (2.15.2)$$

As we know that in the wave propagation there is relation among the velocity v, the wavelength λ and the frequency f as

$$v = \lambda f \qquad (2.15.3)$$

So the velocity can be obtained if the wavelength and frequency are measured. Air is squishy. It can be compressed, as you well know if you ever used a simple bicycle pump.

Sound waves in air are longitudinal waves. A membrane of sound generator (transducer or transmitter) that moves rapidly back and forth making alternate compressions and decompressions of air can create sound waves. When compressions arepassed to the neighboring air particles, they are set in motion because the compressed air nearby pushes on the nearby air particles, much like we visualized for longitudinal waves on a Slinky.

In Fig. 2.15.1, the dots symbolize the air molecules. At a wave crest the air is denser while at a wave trough the air is more dilute. You should imagine the crests and troughs to be moving to the right, assuming a sound generator continues to produce compressions at the left. The picture should not be taken too literally. Obviously, there are many more molecules than we can show, but also the change in air density from one place to the next is tremendously exaggerated.

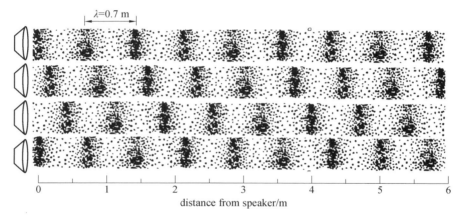

Fig. 2.15.1 Longitudinal sound wave from a generator (loudspeaker). The wave moves left to right and is shown at different instances in time. Greater density of dots indicates greater air pressure. The distance from one wave crest (maximum density) to the next is called the wavelength λ. The bottom drawing is lagging one period T behind the top drawing

The pressure wave (pressure change) without decay can be expressed by

$$p_{\mathrm{f}} = A\cos(\omega t - kx + \varphi_0) \qquad (2.15.4)$$

where A, ω, φ_0 denote the wave amplitude, angular frequency and initial phase respectively; k is the value of wave vector with $k = 2\pi/\lambda$. For convenience we set $\varphi_0 = 0$. We will see what happens when the sound encounters with a receiver (another transducer) which is located at a certain distance away from the transmitter assuming that the receiver is with an end (facet) vertical to the propagation path of ultrasound beam in fact it forms a reflecting end. Hear we assume a plan wave for the sound.

In an ideal case, the reflected wave has the same frequency and amplitude with the incident wave:

$$p_{\mathrm{b}} = A\cos(\omega t + kx + \pi) \qquad (2.15.5)$$

In the expression, a half wave loss (the phase factor of π) and change in propagation direction (the sign " + " in term kx) are considered. The both forward and backward wave interferes and leads to a resultant movement:

$$p = p_f + p_b = A\cos(\omega t - kx) + A\cos(\omega t + kx + \pi)$$
$$= 2A\sin kx \sin \omega t \qquad (2.15.6)$$

In the expression, one can see that the wave amplitude changes with position: in some place the wave vibrates with a maximum of $2A$ while at another place there is no vibration. The still position and the position with maximum vibration are termed as the node and the antinode, respectively, as shown in Fig. 2.15.2. The interval between the adjacent nodes or antinodes is $\lambda/2$, namely the half wavelength.

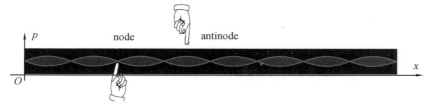

Fig. 2.15.2   The node and antinode

The structure of instrument for measuring sound speed is schematically shown in Fig. 2.15.3. The ultrasonic wave can be emitted via a transmitter and detected by a receiver located at a given distance l from the transmitter, the position of which may be varied by rotating a measuring drum and determined by reading the ruler and the drum. In the experiment, the transmitter is connected to a function generator and accessed to the channel one (CH1), one of the input ports of the oscilloscope, while the receiver to channel two (CH2), another input port of the oscilloscope.

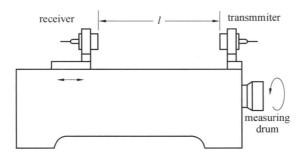

Fig. 2.15.3   The structure of instrument for measuring sound speed

There are several methods to determine the sound speed. In fact, in the experiment, the frequency of the ultrasonic wave is fixed, so one can easily get the sound speed by measuring the wavelength of the wave. We have different ways to measure the wavelength (the half wavelength, more accurately).

(1) The extremum method.

Changing the distance between transmitter and receiver by fixing the position of transmitter, the receiver may obtain a serial of sound pressure signal with peaks and valleys, in which the difference in position of adjacent peaks or valleys is the half wavelength, $\lambda/2$. It is demonstrated theoretically that the peaks and valleys are different in decay rate; the change near maxima is more rapid than near valleys, so usually we choose the measurement for peaks rather than valleys because the peak positioning is easier. In the experiment, one should measure the position for serial of peaks by fixing the frequency of ultrasound near 40 kHz first, then calculate the wavelength, thus the speed of sound can be determined by $v = f\lambda$.

(2) The method of phase comparison.

As we know that at the same time the phase of wave at the transmitter is different from that at the receiver, the phase increment may be acquired by changing the distance between transmitter and receiver. When the distance changes by a half wavelength, $\lambda/2$, the phase changes by one π.

One may respectively input the signals from transmitter and receiver into the input ports for CH1 and CH2 of a two-trace oscilloscope and observe the change in the Lissajous figure with the changing the position of receiver. It is well known that the signal from transmitter and receiver has the same frequency so the Lissajous figure will experience change from ellipses (sometimes circle) with different ellipticity to straight line then back to ellipses. The concrete shape is determined by the phase difference between transmitter and receiver. In fact the positions of receiver for straight lines are most sensitive for the measurement and the difference of adjacent positions corresponding to two straight lines, which have the opposite slopes, is just a half wavelength. So by recording the position for straight lines continually, the positions for a serial of half wavelengths, the sound velocity can be calculated.

(3) The method of waveform movement.

One may indicate simultaneously the waveforms of signals from two channels (one from transmitter another from receiver) in the screen. The change in receiver's position may cause the phase change, which is directly linked to the distance between the transmitter and the receiver and lead to a relative movement of the two waveforms. One can record serial positions corresponding to the states of "in-phase" and "anti-phase". Since the adjacent distance of the two states is just a half wavelength, $\lambda/2$.

3. Experiment contents

(1) Preparation work.

①Connect the various kinds of equipment using the cables and warm up them (2 min at least), link the transmitter and receiver respectively to the CH1 port and CH2 port for the input of the double-traced oscilloscope, adjust the state of the oscilloscope and let the two channels display adequately and independently.

②Be familiar with the usage of oscilloscope, adjust the output of the signal from a function generator and observe their variation.

③Adjust the signal frequency near 40 kHz of the functional generator, move the receiver in the same time until the signal reach the steady maximum. Search the resonance point in frequency for the transducer and record the value of frequency from the function generator display.

(2) Measure the sound speed using the extremum method.

Observe the change in waveform of the output signal from the receiver with respect to that from the emitter with the movement of receiver, continually search the maximum in amplitude and record corresponding positions of receiver.

(3) Measure the sound speed using the method of phase comparison.

Using the X-Y function of oscilloscope to composite the Lissajous figures, continually search and record the positions for straight lines with negative and positive slope (why straight lines?).

(4) Measure the sound speed using the method of waveform movement.

Continually change the receiver's position and record the serial positions corresponding to the states of "in-phase" and "anti-phase".

(5) Phenomenon observation.

Observe the signal change when insert a piece of material such as plastics or cloth into the path between the receiver and emitter, insert a pen and tell what happens.

4. Caution

In the experiment, the accuracy of reading data determines the degree of accuracy of the measured quantity (say the sound speed), so the return difference caused by rotating the measuring drum should be avoided by keeping the direction!

5. Discussion

(1) What is the main factor from the various error sources for measurement of sound speed?

(2) Which one is more sensitive from the used methods?

Experimental Report

Measurement of Sound Velocity in Air

Name: Student ID No.:   Score:

Date: Partner: Teacher's signature:

1. Purpose

2. The experimental data

The resonance frequency for the transducer $f =$ _____ (Hz).

(1) For extremum method (Table 2.15.1).

Table 2.15.1 The positions of series of maxima

Maximum No.	1	2	3	4	5	6	7	8
Receiver position								
Maximum No.	9	10	11	12	13	14	15	16
Receiver position								

(2) For method of phase comparison (method of Lissajous figure) (Table 2.15.2).

Table 2.15.2 The positions of series of straight lines

Line No.	1	2	3	4	5	6	7	8
Line's direction	/	\	/	\	/	\	/	\
Receiver position								
Line No.	9	10	11	12	13	14	15	16
Line's direction	/	\	/	\	/	\	/	\
Receiver position								

(3) For method of waveform movement (Table 2.15.3).

Table 2.15.3 The positions of series of straight lines

No.	1	2	3	4	5	6	7	8
State (in- or anti-phase)	in-	anti-	in-	anti-	in-	anti-	in-	anti-
Receiver position								
No.	9	10	11	12	13	14	15	16
State (in- or anti-phase)	in-	anti-	in-	anti-	in-	anti-	in-	anti-
Receiver position								

3. Data processing

(1) Calculate the standard sound speed.

Environment temperature: $t =$ _____ (℃).

$$v = v_0 \sqrt{\frac{T}{273.15}} = 331.45 \sqrt{1 + \frac{t}{273.15}} =$$

(2) Calculate the wavelength and sound velocity using the successive difference method.

①For maximum method.

$\lambda =$ _____

Sound speed $v = f\lambda =$ _____

②For method of phase comparison (method of Lissajous figure).

$\lambda =$ _____

Sound speed $v = f\lambda =$ _____

③For method of waveform movement.

$\lambda =$ _____

Sound speed $v = f\lambda =$ _____

(3) Compare the results from three methods.

Relative difference:

4. Discussion

2.16 Voltmeter-Ammeter Method for Measurement of Resistance

1. Purpose

(1) Master a method to compute resistance;
(2) Know how to use some electrical instruments;
(3) Learn to analyze and eliminate systematic error.

2. Principle

(1) Ohm's Law.

If a voltage U is applied across an element in an electrical circuit, the current I in the element is determined by a quantity known as the resistance R. The relationship between these three quantities serves as a definition of resistance:

$$R = \frac{U}{I} \qquad (2.16.1)$$

The units of resistance are volt/ampere. However, the combined unit is called the Ohm. The symbol for Ohm is Ω. Some circuit elements obey a relationship known as Ohm's Law. For these elements, the quantity R is a constant for different values of U. If a circuit element obeys <u>Ohm's Law</u>, when the voltage U is varied the current I will also vary, but the ratio U/I should remain constant. A resistor that has constant resistance is also said to be "Ohmic". The plot of U versus I is a straight line (Fig. 2.16.1(a)). Materials that do not obey Ohm's Law are said to be "nonohmic" and have a nonlinear voltage – current relationship. Semiconductors and transistors are nonohmic. For instance, Fig. 2.16.1(b) shows the voltage-current graph of light emitting diode (LED) (which is a semiconductor light source that emits light when current flows through it.). It is not a straight line, which means its resistance is not constant and varies with its voltage.

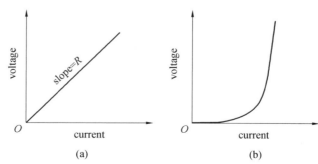

Fig. 2.16.1 Voltage – current graph for an ohmic resistance (a) and for a nonohmic resistance (b)

In an electrical circuit with two or more resistances and a single voltage source, Ohm's Law may be applied to the entire circuit or to any portion of the circuit. When it is applied to the entire circuit, the voltage is the terminal input voltage supplied by the voltage source, and the resistance is the total resistance of the circuit. When Ohm's Law is applied to a particular portion of the circuit, the individual voltage drops, currents, and resistances are used for that part of the circuit.

(2) Arrangements of voltmeter and ammeter.

If we want to know the resistance of a current element, according to Ohm's Law we just need to know its voltage drop measured by voltmeter and current measured by ammeter.

There are two basic arrangements by which resistance is measured with an ammeter and a voltmeter. One circuit is shown in Fig. 2.16.2(a). The current I through the resistance R is measured with an ammeter, and the potential difference or voltage drop U across the resistance is measured with a voltmeter. Then, by Ohm's law, $R = U/I$. Strictly speaking, however, this value calculated by U/I is not correct, since the current registered on the ammeter divides between the resistance R and the voltmeter in parallel. Although a voltmeter is a high-resistance instrument and draws relatively little current, its resistance also cause error (this error is a kind of systematic error). For more accurate resistance measurement, one must take the resistance of the voltmeter into account. The current drawn by the voltmeter is I_V.

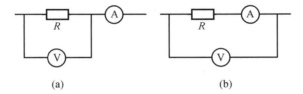

Fig. 2.16.2   The basic arrangements for measuring resistance with
an ammeter and a voltmeter

Since the total current I divides between the resistance and the voltmeter in the parallel branch, there is

$$I = I_R + I_V \tag{2.16.2}$$

where I_R is the true current through the resistance. Then, by Ohm's law, there is

$$R = \frac{U}{I_R} = \frac{U}{I - I_V} = \frac{U}{I - \dfrac{U}{R_V}} \tag{2.16.3}$$

If the voltmeter resistance R_V is much greater than R, to a good approximation, it could be written by

$$R \approx \frac{U}{I} \text{ if } R_V \gg R \tag{2.16.4}$$

which means the measured value is smaller than true value.

Another possible arrangement for measuring R is shown in the circuit diagram in Fig. 2.16.2(b). In this case, the ammeter measures the current through R alone, but now the voltmeter reads the voltage drop across both the ammeter and the resistance. Taking the voltage drop or the resistance of the ammeter into account, then

$$U = U_R + U_A = IR + IR_A \qquad (2.16.5)$$

where R_A is the resistance of ammeter. It means that

$$\frac{U}{I} = R + R_A \qquad (2.16.6)$$

Solving for R from equation above, one gets

$$R = \frac{U}{I} - R_A \qquad (2.16.7)$$

If the ammeter resistance R_A is much smaller than R, to a good approximation, it could be written by

$$R \approx \frac{U}{I} \text{ if } R_V \ll R \qquad (2.16.8)$$

which means the measured value is greater than true value.

(3) Connection of rheostat in the circuit.

A rheostat is a variable resistor with three terminals, two fixed terminals and a moving middle terminal. The middle terminal is called the wiper that can be tuned in two directions. It is often used in circuit to change voltage or current. So, a rheostat works as a voltage controller (Fig. 2.16.3(a)) or a current controller (Fig. 2.16.3(b)).

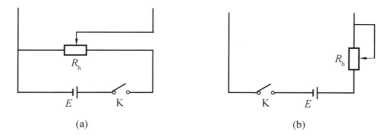

Fig. 2.16.3 Schematic of rheostat connection in circuits. (a) Rheostat works as a voltage controller. (b) Rheostat works as a current controller

R_h—rheostat; K—switch; E—power supply

3. Experimental content

(1) Connect ammeter, voltmeter, rheostat, switch and power supply to resistor as shown in Fig. 2.16.2(a) and Fig. 2.16.3(a).

In our lab, ammeter and voltmeter is digital and have several ranges. Choose appropriate range according to the electromotive force of power supply and the approximate resistance of resistor to be measured (the approximate resistance is supplied by teacher or instructor in the lab). If the ammeter or voltmeter's range is too big, the significant figures is few which lowers the accuracy. If the ammeter or voltmeter's range is too small, the meters are possibly damaged. Also, attention should be given to the proper polarity. Connect " + " of one instrument to " + " of the other instrument.

(2) Check out the circuit. Be sure it is right and then close the switch to active the circuit. Adjust rheostat and record 6 different readings of voltage and current in Table 2.16.1 in the experimental report below.

(3) Using Eq. (2.16.4), compute the resistance for each current setting and find the average value. (Note that the resistance of digital voltmeter in lab is much greater than R to be measured so Eq. (2.16.4) is accurate enough.)

(4) Set up a circuit as shown in Fig. 2.16.2(b) and Fig. 2.16.3(a). Repeat the measurements as in procedure (2) for this circuit, recording readings in Table 2.16.2 in the experimental report below.

(5) In this case, ammeter resistance should be taken into account. Using Eq. (2.16.7), compute the resistance for each current setting and find the average value. (One can get ammeter resistance from teacher or instructor in lab.)

(6) Repeat the previous procedures with another known resistor. Record data in the space provided in Table 2.16.3 and Table 2.16.4 in the experimental report below.

(7) Plot volt-ampere characteristic curve of LED. Set up a circuit as shown in Fig. 2.16.2(b) and Fig. 2.16.3(a) and just replace resistor with a LED. Make sure the current through LED is lower than its maximum forward current which is the maximum value of the forward current that a diode can carry without damaging the device. If the forward current in a diode is increased beyond this specified value, the junction will be destroyed. One can get ammeter resistance from teacher or instructor in lab.

(8) Adjust rheostat and record 12 different readings of voltage and current in Table 2.16.5 in the experimental report below. Plot volt-ampere characteristic curve in coordinate paper.

Experimental Report

Voltmeter-Ammeter Method for Measurement of Resistance

Name: Student ID No.:   Score:

Date: Partner: Teacher's signature:

1. Purpose

2. The experimental data and processing

(1) Measure the first resistor.

①Set up a circuit as shown in Fig. 2.16.2(a) and Fig. 2.16.3(a) (Table 2.16.1).

Table 2.16.1   Readings of voltage and current

U/V					
I/A					

Using Eq.(2.16.4), compute the average of resistance.

②Set up a circuit as shown in Fig. 2.16.2(b) and Fig. 2.16.3(a) (Table 2.16.2).

Table 2.16.2   Readings of voltage and current

U/V					
I/A					

Using Eq.(2.16.7), compute the average of resistance.

(2) Measure the second resistor.

①Set up a circuit as shown in Fig. 2.16.2(a) and Fig. 2.16.3(a) (Table 2.16.3).

Table 2.16.3   Readings of voltage and current

U/V					
I/A					

Using Eq.(2.16.4), compute the average of resistance.

②Set up a circuit as shown in Fig. 2.16.2(b) and Fig. 2.16.3(a) (Table 2.16.4).

Table 2.16.4   Readings of voltage and current

U/V					
I/A					

Using Eq.(2.16.7), compute the average of resistance.

(3) Set up a circuit as shown in Fig. 2.16.2(b) and Fig. 2.16.3(a) and replace resistor with a LED(Table 2.16.5).

Table 2.16.5 Readings of voltage and current

U/V										
I/A										

Plot voltage − current characteristic curve of LED.

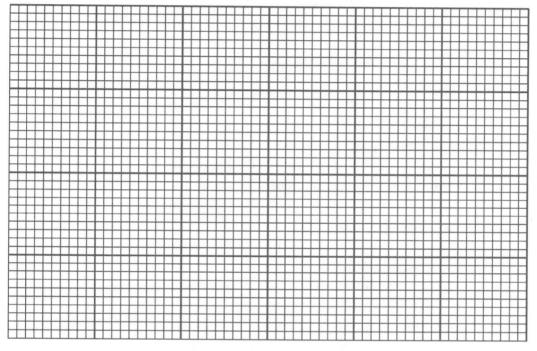

3. Discussion

2.17 Transient Process of a RC Circuit

1. Purpose

(1) To observe the charging and discharging process of a capacitor in RC circuit and measure the time constant for RC circuit;

(2) To observe the waveforms of capacitor and resistance at different time constant on oscilloscope;

(3) To re-practice the usage of oscilloscope.

2. Principle

(1) Transient process of the RC circuit.

Capacitor is a storage appliance of electric charges. The process that the capacitor plates accumulate or release electric charges is called charging or discharging of a capacitor. At any time during the process, the quantity of electricity or charge q is directly related to the voltage across the capacitor plates U_C, namely $q = CU_C$, where the coefficient C refers to the capacitance of the capacitor in Farads (F).

Fig. 2.17.1 illustrates the principle for charging and discharging of a capacitor. When the key K is switched to position 1, the DC power source E starts to charge the capacitor. Meanwhile there is a current change in the circuit and U_C gradually increases as q accumulates. When $U_C = E$ with E being electromotive force of the source, the charging stops. When the key K switches to the position 2, the capacitor begins to discharge. The charge q on the capacitor gradually reduces through resistor and the capacitor voltage U_C decreases with discharging current. The process will not be over until both the quantity of electricity q and capacitor voltage U_C become zero. Actually, the charging or discharging process is a transition process from one steady state to another steady state in the RC circuit, which is stimulated by the change of the step voltage. The process is also known as a transient process. In this process, the current i in the circuit, the charge q on the capacitor and voltage U_C across the capacitor plates present certain regularity, which is determined by the circuit equation and initial conditions.

For the charging process, the following circuit equation is obeyed:

$$U_C + iR = E \quad (2.17.1)$$

where E is the electromotive force of the DC source. The charging current i is related to the changing rate of charge, namely $i = dq/dt$. Due to $q = CU_C$, charging current i can be written as $i = CdU_C/dt$, which depends on the raising rate of the voltage. Thus Eq. (2.17.1) can be converted into

$$U_C + RC\frac{dU_C}{dt} = E \quad (2.17.2)$$

Part 2 Topics of Experiments

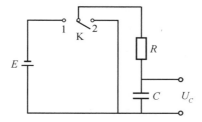

Fig. 2.17.1   The schematic graph for charging and discharging for a capacitor C through a resistor R by a DC power supply E, where K denotes the single-pole and double-throw switch

which is in fact a differential equation with respect to the voltage U_C. In the process of charging, the capacitor starts with no voltage, i.e. $U_C = 0$ at $t = 0$, and at the end of process, voltage reach a maximum, namely $U_C = E$. On the contrary, in the process of discharging, at very beginning $t = 0$, $U_C = E$ and at the end of discharging, $U_C = 0$. Under these initial conditions the solutions of equations are as follow:

$$U_C = E(1 - e^{-t/\tau})$$

$$i = \frac{E}{R} e^{-t/\tau} \text{ (for charging)} \qquad (2.17.3)$$

$$U_C = E e^{-t/\tau}$$

$$i = -\frac{E}{R} e^{-t/\tau} \text{ (for discharging)} \qquad (2.17.4)$$

where $\tau = RC$.

These equations reflect respectively the charging and discharging process. It is obvious that in both processes of charging and discharging the voltage U_C and the current i rise and decay respectively with time t with the same exponential parameter τ. However, the capacitor voltage rises gradually in charging and declines in discharging. The minus sign in Eq.(2.17.3) and Eq.(2.17.4) indicates that the discharging current and the charging current are in the opposite direction.

The characteristics of voltage on capacitor can also be reflected by the voltage U_R across the resistance R. U_R and i have the same changing regularity because $U_R = iR$. Meanwhile, it is known from Eq. (2.17.3) and Eq. (2.17.3) that q and U_C also have the same changing regularity.

(2) The role of time constant.

It is clear that the charging and discharging rate in the process can be characterized by a time constant $\tau = RC$ which is the product of resistance R and the capacitance C. If the capacitor voltage is zero at the beginning of charging and the charging current is I_0, the capacitor can achieve a stable voltage U_0 at the end of charging. From Eq.(2.17.3), one can see that at time $t = \tau = RC$ the voltage across capacitor is

$$U_C = U_0(1 - e^{-1}) \sim 0.63 U_0$$

which means the voltage is about 63% of its maximum. While the current becomes

$$i = I_0 e^{-1} \sim 0.37 I_0$$

which reduces to about 37% of its initial value.

During the discharging process of an RC circuit, from Eq.(2.17.4) if the initial voltage of the capacitor is U_0, after a discharging time of τ, the residual voltage of capacitor is

$$U_C = U_0 e^{-1} \sim 0.37 U_0$$

Obviously, time constant of a RC circuit expresses the rate of charging and discharging. Thus the quantity τ is an important parameter to reflect the characteristics of RC circuit. Usually, during the charging and discharging process of a RC circuit, after five time constants (5τ) the capacitor voltage almost reaches a steady state. Fig. 2.17.2 gives the typical curves (V vs t) for capacitor charging and discharging, from which one can see the role of the time constant.

The time constant τ can be determined through experiments. Taking the natural logarithm (base e) of Eq.(2.17.3) and Eq.(2.17.4), one can get

$$\ln(E - U_C) = \frac{t}{\tau} + \ln E \quad \text{(for charging)} \tag{2.17.5}$$

$$\ln U_C = -\frac{t}{\tau} + \ln E \quad \text{(for discharging)} \tag{2.17.6}$$

Thus, by calculating the slope of curves of $\ln(E - U_C) - t$ or $\ln U_C - t$, one could obtain the time constant τ.

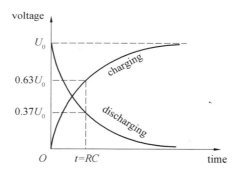

Fig. 2.17.2 A graph illustrating voltage versus time for capacitor charging and discharging. The "steepness" of the curve depends on the time constant RC

(3) Steady waveform of RC circuit stimulated by square wave.

Rectangular waveform is also called square waveform (Fig. 2.17.3), a common series of pulses in the electronic circuit. If the RC circuit is stimulated by square wave, the capacitor will experience alternative charge and discharge. After some transient processes, steady waveform, i.e. capacitor voltage U_C and resistance voltage U_R can be formed. The steady-state waveform has the following characteristics:

(1) Steady-state waves are symmetric about t-axis;

(2) Corresponding to the rising edge of square wave, U_C is negative and corresponding to

the falling edge of square wave, U_C is positive.

(3) At any time, there is

$$U_C + U_R = U_P \tag{2.17.7}$$

where U_P is the peak voltage of square wave.

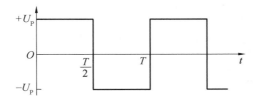

Fig. 2.17.3 Square waveform. Its peak voltage is U_P and period is T

Equations below are the changing regularity of capacitor voltage and resistance voltage with time in the steady state within half period:

$$U_C(t) = U_P - (U_P + U_0) e^{-t/\tau} \tag{2.17.8}$$

$$U_R(t) = (U_P + U_0) e^{-t/\tau} \tag{2.17.9}$$

where U_0 denotes an instantaneous value of U_C at the square wave edge. When $t = T/2$, $U_C(t) = U$, Eq.(2.17.8) is rewritten as

$$U_0 = \frac{1 - e^{-T/2\tau}}{1 + e^{-T/2\tau}} U_P \tag{2.17.10}$$

It can be seen that the value of U_0 and also U_C and U_R depend on the ratio of the period T of the rectangular wave to time constant τ. Because the steady-state waves are symmetric about t-axis, the steady state waves can be drawn by Eq.(2.17.8) and Eq.(2.17.9). The waveforms of U_C and U_R are shown in Fig. 2.17.4(a) and Fig. 2.17.4(b) respectively when $\tau > T/2$. In this case, at $t = T/2$, because τ is large, capacitor voltage doesn't reach saturation value U_P but reach U_0 and then capacitor begins to reverse charge until U_C achieves the maximum $-U_0$. When $\tau < T/2$, because τ is small, U_C quickly reaches the saturation value U_P (Fig. 2.17.5(a)) and the waveform of U_C approximates to square wave. The waveform of U_R, which is the same as the circuit current waveform, presents sharp pulse (Fig. 2.17.5(b)).

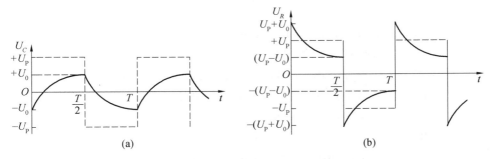

Fig. 2.17.4 Steady state wave pattern under square wave when $\tau > T/2$. (a) is the voltage of capacitor and (b) is the voltage of resistance

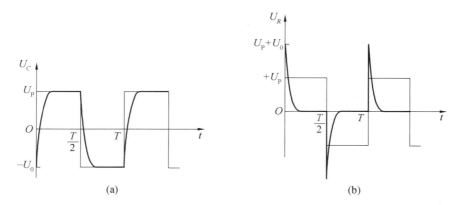

Fig. 2.17.5　Steady state wave pattern under square wave when $\tau < T/2$. (a) is the voltage of capacitor and (b) is the voltage of resistance

3. Experimental content

(1) Connect circuit according to Fig. 2.17.6(a). The circuit is a measurement circuit for the discharging curve and discharging time constant. DVM is a digital voltmeter whose resistance is R_V and R_1 is the resistor whose resistance is 1 000 Ω. Pay attention to the polarity of capacitor because C is an electrolytic capacitance. Set electric force 6 V and close K_1 to charge the capacitor.

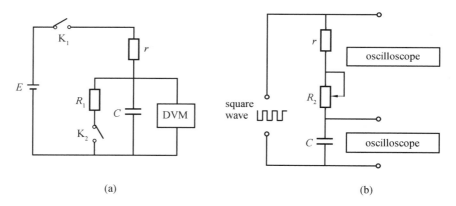

Fig. 2.17.6　(a) Circuit for discharging curve and discharging time constant. (b) Circuit for observing the waveform by oscilloscope

①Based on Fig. 2.17.6(a), disconnect the electric source (open K_1). Open the switch K_2 in order to make the capacitor discharge through DVM. Meanwhile record $U_C(t)$ every 20 s and fill in the blanks in Table 2.17.1 in the experimental report.

②Connect the electric source (close K_1) to make the capacitor charge again. Few seconds later disconnect the electric source (open K_1) and at the same time close K_2 to make the capacitor discharge through DVM and R_1. Record $U_C(t')$ every 20 s and fill in the blanks in Table 2.17.2 in the experimental report.

③Plot discharging curve $\ln U_C(t) - t$ and $\ln U_C(t') - t'$ respectively on your coordinate paper and calculate C from two slopes.

(2) Connect circuit according to Fig. 2.17.6(b) in which R_2 is continuously variable resistor. Choose square wave whose period is 0.02 s and observe the following waveforms by oscilloscope:

①The shape of square wave;
②Waveforms of U_C and U_R respectively when R_2 are the maximum;
③Waveforms of U_C and U_R respectively when R_2 are the minimum.

Plot these waveforms on your report.

Experimental Report

Transient Process of a RC Circuit

Name: Student ID No.: Score:

Date: Partner: Teacher's signature:

1. Purpose

2. The experimental data and processing

(1) Connect circuit according to Fig. 2.17.6(a).

①Discharging with K_2 open (Table 2.17.1).

Table 2.17.1 Voltages of capacitor every 20 s

t/s	0	20	40	60	80	100	120
$U_C(t)/V$							

②Discharging with K_2 close (Table 2.17.2).

Table 2.17.2 Voltages of capacitor every 20 s

t'/s	0	20	40	60	80	100	120
$U_C(t')/V$							

Plot discharging curves $\ln U_C(t) - t$ and $\ln U_C(t') - t'$ respectively on the same graph paper.

For each curve, select two points to compute the slope.

(2) Connect circuit according to Fig. 2.17.6(b).

Observe waveforms of U_C and U_R through oscilloscope when R_2 are the maximum and minimum, respectively. Plot these waveforms below.

Part 2 Topics of Experiments

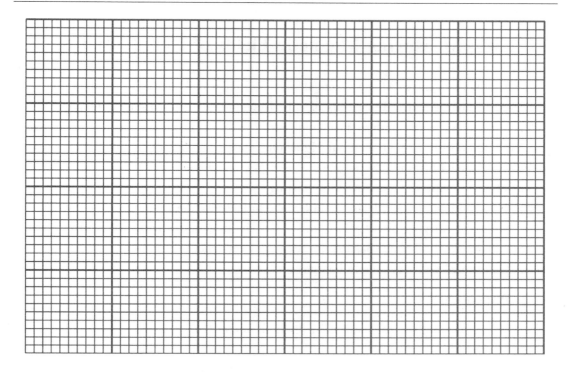

3. Discussion

2.18 Measuring Magnetic Fields of Helmholtz Coil

Helmholtz coil consists of two identical coaxial circular coils that are separated from each other by a distance equal to the radius of coil. Helmholtz coil has been extensively used to produce a region of nearly uniform magnetic field. Due to its advantages such as large region of uniform field with free of material, Helmholtz coils are normally applied for magnetic calibration, cancelling out background (earth's) magnetic field, and for electronic equipment of magnetic field susceptibility testing.

In this experiment, we will use a Hall probe to measure the magnetic field distributions of both a current-carrying circular coil and a Helmholtz coil. Utilizing the Hall effect of a semiconductor, the Hall probe has the advantages of very small volume, high accuracy, wide measurement range and very simple instrumentation.

1. Purpose

(1) To understand the magnetic field distributions in both current-carrying circular coil and Helmholtz coil;

(2) To master how to use the Hall probe for measuring the low magnetic field;

(3) To verify the principle of superposition of fields.

2. Principle

(1) The magnetic field of a current-carrying circular coil.

Derived from the Biot-Savart law, the on-axis magnetic field due to a circular coil with current I, radius R and number of turns N, has the expression

$$B = \frac{\mu_0 N R^2 I}{2(R^2 + x^2)^{3/2}} \qquad (2.18.1)$$

where μ_0 is vacuum permeability ($\mu_0 = 4\pi \times 10^{-7}$ H/m) and x is the axial distance from the center of the coil. The direction of the magnetic field is along the coil's axis. Fig. 2.18.1 shows the schematic drawings of single circular coil and magnetic field distribution along its axis.

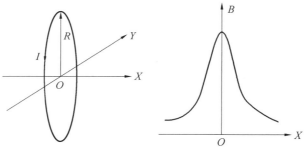

Fig. 2.18.1 Schematic drawings of single circular coil (left) and magnetic field distribution along its axis (right)

Part 2 Topics of Experiments

(2) The magnetic field of Helmholtz coil.

A Helmholtz coil consists of two identical circular current coils of radius R, each having N turns. The two coils are parallel and coaxial, and separated by distance R, as shown in Fig. 2.18.2.

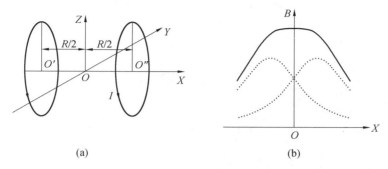

(a)   (b)

Fig. 2.18.2 (a) Geometry of the Helmholtz coil, here X axis is the coil axis. (b) Schematic drawings of magnetic field distribution on axis of Helmholtz coil (solid line) and every single coil (dashed line)

Each coil carries an identical electric current I in the same direction. We choose the x axis along the axis of the coils, and choose the origin at the midpoint between the coils, as shown in Fig. 2.18.2. By the principle of superposition of fields, the total field of the Helmholtz coil is the sum of the fields from each coil. While, each coil's field can be easily obtained from the expression (Eq. (2.18.1)) for the single circular coil by applying the coordinate transformations $x \rightarrow x - R/2$ and $x \rightarrow x + R/2$ for the right coil, and for the left coil, respectively. Thus, the total axial field is given by

$$B = \frac{\mu_0 N R^2 I}{2} \left[\frac{1}{\left[R^2 + \left(x + \frac{R}{2}\right)^2\right]^{3/2}} + \frac{1}{\left[R^2 + \left(x - \frac{R}{2}\right)^2\right]^{3/2}} \right] \quad (2.18.2)$$

In the region midway between two coils, the magnetic field distribution is almost uniform (see the schematic drawing in Fig. 2.18.2(b)). By setting $x = 0$ in Eq. (2.18.2), we can obtain the magnetic field at the midpoint between the coils:

$$B(0) = \frac{8}{5^{3/2}} \frac{\mu_0 N I}{R} \quad (2.18.3)$$

Helmholtz coils are frequently used in the laboratory because they provide a reasonably large region free of material in which there is a uniform magnetic field (the region midway between the coils and along the axis). Besides making magnetic fields, Helmholtz coil is also used to cancel out the Earth's magnetic field in some applications. Notably, although most Helmholtz coils are circular, some are square. This is because that square Helmholtz coil offers large working space than circular.

3. Apparatus

The schematic drawing of the apparatus is shown in Fig. 2.18.3. It mainly consists of three parts: ① two circular coils and measurement plate having a one centimeter grid; ② an instrument box having high sensitivity 3.5 digit millitesla meter, DC stabilized current supply and digital ammeter; ③ 95 A integrated Hall probe.

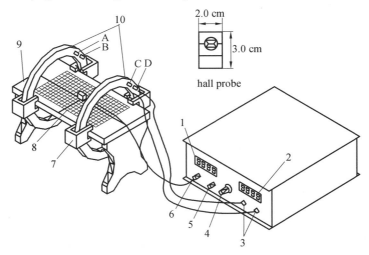

Fig. 2.18.3 The schematic drawing of the apparatus
1—millitesla meter; 2—ammeter; 3—DC stabilized current supply; 4—current control knob; 5—zero control knob; 6—plug for probe; 7—holder; 8—Hall probe; 9—measurement plate; 10—two circular coils; A, B, C, D—terminals

4. Experiment contents

(1) Measure single current-carrying circular coil's magnetic field along its axis from $x = -12.0$ cm to $x = 12.0$ cm in steps of 1.0 cm. Compare the experimental results with the theoretical ones calculated with Eq.(2.18.1).

(2) For Helmholtz coil (the separation between the two coils is equal to the radius of coil $R = 10.00$ cm), first measure the fields produced by each coil separately, then connect the two coils in series and measure the field of Helmholtz coil. All measurements are carried on the coil's axis from $x = -12.0$ cm to $x = 12.0$ cm at intervals of 1.0 cm.

(3) Change the separation between the two coils and measure the field along the axis of coil from $x = -12.0$ cm to $x = 12.0$ cm in steps of 1.0 cm. (Suggestion: coil spacing $d = R/2 = 5.0$ cm and $d = 2R = 20.0$ cm.)

(4) With the Hall probe, measure the Earth's magnetic field.

Relative parameters: current $I = 100$ mA, radius of coil $R = 10.00$ cm, number of coil's turns $N = 500$.

5. Cautions

(1) Adjust the holders to make sure that the each coil's axis lies exactly at the middle line of the measurement plate.

(2) Before making measurement of each field point, zero the Hall probe.

Place the Hall probe at the field point, disconnect the current and rotate the zero contact knob until magnetometer is zeroed. The purpose of zeroing the probe is to cancel out the effect of background (earth's) magnetic field.

(3) Pay attention to the direction of current in coils. Use the red terminal attached to the coil as the positive terminal.

6. Discussion

(1) When the current through the two coils flows in the opposite direction, what is the magnetic field distribution along its axis in Helmholtz coils?

(2) What is the procedures in measuring the Earth's magnetic field with Hall probe?

Experimental Report

Measuring Magnetic Fields of Helmholtz Coil

Name:　　　　　　　　Student ID No.:　　　　　　　Score:

Date:　　　　　　　　　Partner:　　　　　　　　　　Teacher's signature:

1. Purpose

2. Data recording and processing

Relative parameters: current $I = $ _____ mA, radius of coil $R = 10.00$ cm, number of coil's turns $N = 500$.

(1) Fulfill Table 2.18.1.

Table 2.18.1　Single circular coil's magnetic field along its axis

x/cm	B_{exp}/mT	B_{theo}/mT	x/cm	B_{exp}/mT	B_{theo}/mT
−12.0			1.0		
−11.0			2.0		
−10.0			3.0		
−9.0			4.0		
−8.0			5.0		
−7.0			6.0		
−6.0			7.0		
−5.0			8.0		
−4.0			9.0		
−3.0			10.0		
−2.0			11.0		
−1.0			12.0		
0.0					

Note: B_{exp} and B_{theo} denote the measured results and the theoretical results respectively.

(2) Plot the curves of B_{exp} and B_{theo} versus x on graph paper.

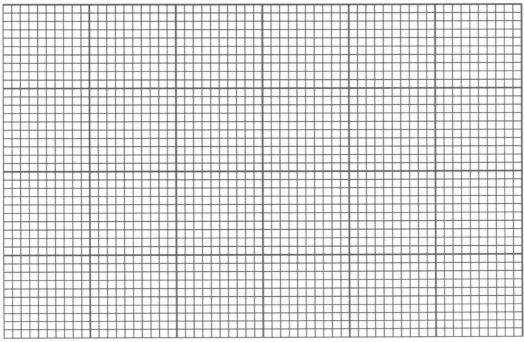

(3) Fulfill Table 2.18.2.

Table 2.18.2 The magnetic field on axis of Helmholtz coil (coil spacing 10.0 cm)

x/cm	B_L/mT	B_R/mT	B_{L+R}/mT	$(B_L + B_R)$/mT
−12.0				
−11.0				
−10.0				
−9.0				
−8.0				
−7.0				
−6.0				
−5.0				
−4.0				
−3.0				
−2.0				
−1.0				
0.0				
1.0				
2.0				

Continued   Table   2.18.2

x/cm	B_L/mT	B_R/mT	B_{L+R}/mT	$(B_L + B_R)$/mT
3.0				
4.0				
5.0				
6.0				
7.0				
8.0				
9.0				
10.0				
11.0				
12.0				

Here, B_L and B_R represent for the fields produced by the left coil and right coil respectively. B_{L+R} denotes the results of Helmholtz coil.

(4) Plot the curves of B_L, B_R, B_{L+R} and $B_L + B_R$ as functions of x on graph paper.

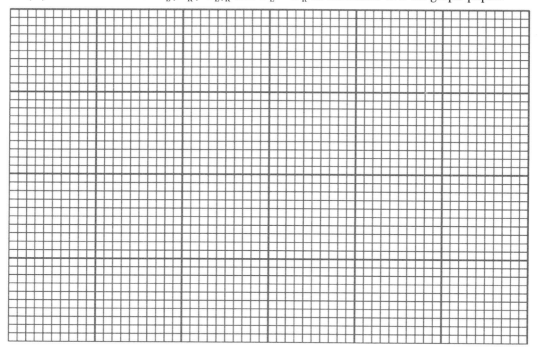

(5) Fulfill Table 2.18.3.

Table 2.18.3 The magnetic fields on axis produced by Helmholtz pairs with coil spacing $d = R/2 = 5.0$ cm, denoted as B_5, and $d = 2R = 20.0$ cm, denoted as B_{20}

x/cm	B_5/mT	B_{20}/mT	x/cm	B_5/mT	B_{20}/mT
−12.0			1.0		
−11.0			2.0		
−10.0			3.0		
−9.0			4.0		
−8.0			5.0		
−7.0			6.0		
−6.0			7.0		
−5.0			8.0		
−4.0			9.0		
−3.0			10.0		
−2.0			11.0		
−1.0			12.0		
0.0					

(6) Plot the curves of B_5 and B_{20} versus x on graph paper.

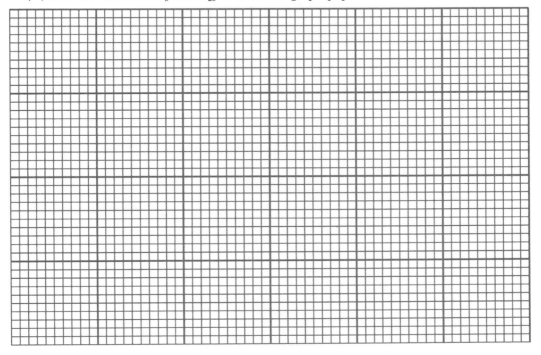

3. Conclusion

4. Discussion

2.19 Dynamic Hysteresis Loop of Ferromagnetic Material

Ferromagnetic materials have been extensively used in various areas from scientific research to technical application. Hysteresis loop and magnetization curve are main characteristics of ferromagnetic materials and also the main criteria in design and fabrication of magnetic components.

In this experiment, we display and measure the dynamic magnetic loop with oscilloscope.

1. Purpose

(1) To deeply understand the definitions of hysteresis loop, magnetization curve, remanence and coercivity;

(2) To master how to use oscilloscope for displaying and measuring the hysteresis loop of the ferromagnetic material.

2. Principle

(1) Ferromagnetic materials and hysteresis loop.

The ferromagnetic materials, such as iron, nickel, and cobalt, are those substances which exhibit strong magnetism in the same direction of the field, when they are placed in magnetic field. Moreover, such materials are able to retain partially magnetized even after the external field is removed. Their powerful influence on magnetization is due to the existence of magnetized domains. Domains are microscopic regions within which all of the magnetic dipoles are aligned. In the absence of an external magnetic field, the individual domains in an unmagnetized material have a random orientation as shown in Fig. 2.19.1(a). Thus, the net magnetization in the material as a whole is zero. When we place an unmagnetized sample of a ferromagnetic material into an external magnetic field, the domains will align with the external field. The alignment may be partial when the field is relative weak, as illustrated in Fig. 2.19.1(b), or even full when the field is strong enough to overcome the resistance, as shown in Fig. 2.19.1(c).

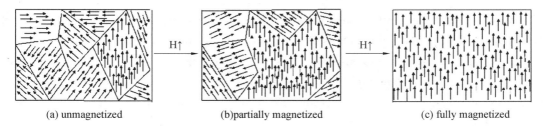

(a) unmagnetized   (b) partially magnetized   (c) fully magnetized

Fig. 2.19.1　Schematic drawing of unmagnetized, partially magnetized, and fully magnetized domains in a ferromagnetic material

H—the external field and vertical arrow;　↑—the direction of field

The magnetization behavior of a ferromagnetic material is described in terms of its $B - H$ magnetization curve, where B and H refer to the amplitudes of the magnetic flux density and magnetic field strength respectively. When a magnetic field is applied to an unmagnetized sample of ferromagnetic material, magnetization initially increases slowly, then more rapidly as the domains begin to grow. Later, magnetization slows, as domains eventually rotate to reach saturation. This magnetization process is characterized by the notion of initial magnetization curve (i.e., the sketched $B - H$ curve from point O to point a in Fig. 2.19.2). At point "$\underline{a}$", representing a saturation condition, nearly all the domains have become aligned with H. If H is now reduced monotonically to zero, the $B - H$ relation does not turn back along the $\underline{O - a}$ curve in Fig. 2.19.2, but instead follows a different curve (i.e., $a - b$ curve as shown in Fig. 2.19.2). At point "b", a residual magnetization, known as the remanence, B_r, is present in the material even though the external field H is zero. The material now acts as a permanent magnet owing to the fact that many of its magnetized domains have remained oriented near the direction of the original field. If we now reverse the direction of H and increasing its intensity, magnetic flux density B decrease from B_r, at point "b", to zero, at point "c". Thus, a coercivity H_c is required to remove the residual magnetism from the material. Further increases in the strength of the field eventually causes B to be saturated but in the opposite direction (point "d").

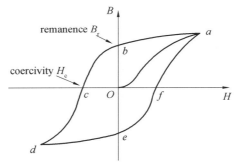

Fig. 2.19.2   The initial magnetization curve (solid line $\underline{O - a}$) and hysteresis loop (solid line $\underline{a \to b \to c \to d \to e \to f \to a}$)

As the field continually alternates, the $B - H$ curve traces out a hysteresis loop. Therefore, if the applied field is an oscillating field, the ferromagnetic material is continually cycled through the same hysteresis loop. This $B - H$ loop is referred as dynamic hysteresis loop. When we increase the amplitude of the alternating field from zero, the tips of the set of hysteresis loops trace out a curve which is named as fundamental magnetization curve (Fig. 2.19.3).

Generally, before measurement of initial magnetization curve or fundamental magnetization curve, the initial state of the sample material is required to be unmagnetized ($H = 0$, $B = 0$). One way to demagnetize a magnetized ferromagnetic material is to place it into a coil. At first, we apply alternating current with large amplitude into the coil so that the ferromagnetic material is under saturation. Then we gradually decrease the amplitude of the alternating current until to zero and achieve the demagnetization.

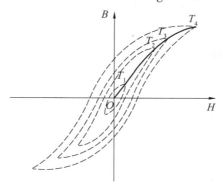

Fig. 2.19.3   Fundamental magnetization curve

$0 \to T_1 \to T_2 \to T_3 \to T_4$

(2) The principle of displaying the dynamic hysteresis loop with oscilloscope.

In this experiment, we use an oscilloscope operating in the Y versus X mode to display dynamic hysteresis loop. Fig. 2.19.4 shows the principle diagram of an excitation and test circuit used to display the dynamic hysteresis loop of a core material upon the screen of oscilloscope. The iron core contains both primary (N turns) and secondary (n turns) winding. The resistor R_1 is in series with primary winding and is used as a current sensing element. The voltage drop across the resistor R_1 is applied to the input channel X (CH1) of oscilloscope to provide a deflection voltage to the horizontal deflecting plates of the oscilloscope. The secondary winding is in series with the resistor R_2 and the output capacitor C. The output of capacitor C is applied to the input channel Y (CH2) of oscilloscope to supply voltage to the vertical deflecting plates.

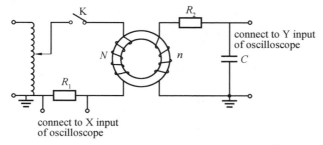

Fig. 2.19.4   Circuit diagram of experimental setup used to display dynamic hysteresis loop of a ferromagnetic material with oscilloscope

Next, we analyze why this circuit system can display and measure the dynamic hysteresis loop.

①The magnetic field strength H is proportional to the output voltage U_{R_1} of resistor R_1. If an oscillating current i_1 passes through the primary winding of N turns, it produces magnetic field H:

$$H = \frac{N}{l} i_1 \qquad (2.19.1)$$

where l is the mean magnetic path length. Notably, in Eq. (2.19.1), we have ignored the impact of the secondary winding current on H. Since the primary winding and R_1 are in series, the same current i_1 passes through them. Substituting $i_1 = U_{R_1}/R_1$, obtained from Ohm's Law, into Eq.(2.19.1), we obtain

$$H = \frac{N U_{R_1}}{l R_1} \qquad (2.19.2)$$

Obviously, for given N, l and R_1, H is proportional to voltage U_{R_1} applied to the input channel X (CH1) of oscilloscope.

②Magnetic flux density B is proportional to the terminal voltage U_C of capacitor C. According to Faraday's law, the induced oscillating voltage e_2 at secondary terminals has the form:

$$e_2 = - nA \frac{dB}{dt} \qquad (2.19.3)$$

where n is the turns of secondary winding; A is the cross-sectional area of rectangular iron core. Since the resistance of R_2 is far larger than that of the capacitor C, we have

$$i_2 = \frac{U_{R_2}}{R_2} = -\frac{e_2}{R_2} \qquad (2.19.4)$$

where i_2 is secondary current; U_{R_2} is the voltage drop across resistor R_2. Meanwhile, there is

$$i_2 = C \frac{dU_C}{dt} \qquad (2.19.5)$$

where U_C is the terminal voltage of capacitor C.

From Eq.(2.19.3) – (2.19.5), we obtain

$$dB = \frac{R_2 C}{nA} dU_C \qquad (2.19.6)$$

Finally, integrating Eq.(2.19.6), we have

$$B = \frac{R_2 C}{nA} U_C \qquad (2.19.7)$$

Thus, for given R_2, C, A and n, the magnetic flux density B is proportional to the voltage U_C of capacitor C, which is applied to the input channel Y (CH2) of oscilloscope. In summary, obviously the Lissajous figure generated by the circuit system on the oscilloscope's screen is indicative of the shape of the dynamic hysteresis loop of the ferromagnetic material under test.

In order to minimize the measurement error introduced by the setup shown in Fig. 2.19.4, the following points should be considered: (a) the current sensing resistor R_1 is required to have a sufficiently small resistance so that the voltage wave form applied to the primary winding closely matched that of the excitation source; (b) the resistor R_2 in secondary circuit should be large enough to make sure that the secondary current referred to the primary winding, $(n/N)i_2$, is negligibly small with respect to the primary current i_1; (c) the circuit R_2C is an integrating circuit and the integrator time constant R_2C is needed to be much larger than the period of the excitation source. With the proper selection of component values, especially the values of R_1 and R_2 in our experiment, we can reduced the measurement errors and / or the distortions of loop display.

(3) Measurement of hysteresis loop.

In order to quantitatively measure the hysteresis loop displayed on the oscilloscope screen, we should calibrate the vertical sensitivity of both the Y and X channels. Generally, oscilloscope provides calibration terminal generating a sample square wave. We can connect this terminal to the input of X (CH1) or Y (CH2) channel to calibrate the oscilloscope's display. Perhaps, we can directly use the vertical sensitivity control, labeled VOLTS/DIV, to set the vertical scale factor (be sure to turn off the fine adjustment knob in the center of the vertical sensitivity control, i.e. rotate the fine adjustment knob fully clockwise). In digital oscilloscope, the vertical sensitivity of each channel has been shown on the screen. Be careful that, after calibration, we cannot further adjust the vertical sensitivity control during the measurement.

Assume that the vertical sensitivities of CH1 (X input) and CH2 (Y input) are D_x (VOLTS/DIV) and D_y (VOLTS/DIV) respectively, and that the coordinates of one point on the loop are x (div) and y (div), we have

$$U_{R_1} = U_x = D_x \cdot x \tag{2.19.8}$$

$$U_C = U_y = D_y \cdot y \tag{2.19.9}$$

Substituting Eq. (2.19.8) and Eq. (2.19.9) into Eq. (2.19.2) and Eq. (2.19.7) respectively, we obtain

$$H = \frac{ND_x}{lR_1}x \tag{2.19.10}$$

$$B = \frac{R_2CD_y}{nA}y \tag{2.19.11}$$

Therefore, we can calculate the values of H and B for each point on the displayed loop and then plot the hysteresis loop.

3. Apparatus

Primary winding (500 turns), secondary winding (1 000 turns), a rectangular iron core, oscilloscope, resistance box (R_1), fixed resistor (10 kΩ, R_2), capacitor (10 μF), step down transformer, autotransformer, vernier caliper, connection wire, etc.

4. Experiment contents

(1) Preliminaries. ①Hook up the circuit shown in Fig. 2.19.4. ②Set the oscilloscope in

XY operation. ③Set the AC-GND-DC Switch in GND to disconnect the input signals of both channels, and adjust the horizontal and vertical position controls in order that the light spot lies exactly at the middle of the screen. ④Connect the input signals again by setting the AC-GND-DC Switch in DC. ⑤If the two tips of the loop lie at quadrant II and quadrant IV, please invert the input signal applied in CH2. ⑥If necessary, select the proper value of R_1 and adjust the VOLTS/DIV controls until the area of the displayed saturation hysteresis loop is as large as possible. ⑦Demagnetize the sample by decreasing the output of the autotransformer from maximum until to zero.

(2) Measure the fundamental magnetization curve. Raise the output of the autotransformer from zero, and record the coordinates of the tips in the first quadrant of the loops, until saturation.

(3) Measure the saturation hysteresis loop. Adjust the vertical sensitivities of CH1 (X input) and CH2 (Y input) and turn the output of the autotransformer in order to obtain a large saturation hysteresis loop but without any distortion. Must record the following points on the loop: $(0, \pm B_r)$, $(H_c, 0)$, and two tips.

(4) Record the vertical sensitivities D_x (VOLTS/DIV) and D_y (VOLTS/DIV).

(5) Measure (using a vernier caliper) the dimensions of the core material.

(6) Calculate the corresponding values of H and B, using Eq.(2.19.10) and Eq.(2.19.11).

(7) Plot the fundamental magnetization curve and the saturation hysteresis loop on graph paper in the same coordinate system.

(8) Provide the values of the remanence B_r and the coercive field H_c.

5. Cautions

(1) Before measurement: ①set the light spot at the middle of the screen when disconnecting the input signals of both two channels; ②demagnetize the sample by decreasing the output of the autotransformer from maximum until to zero.

(2) With the proper selections of the resistor R_1, the vertical sensitivities and the output of autotransformer, the displayed saturation hysteresis loop should be as large as possible and without any distortion.

(3) During the measurement, don't further adjust the vertical sensitivity controls.

6. Discussion

(1) How to demagnetize a ferromagnetic material?

(2) What are the different types of ferromagnetic materials? What are their properties and applications?

Part 2　Topics of Experiments

Experimental Report

Dynamic Hysteresis Loop of Ferromagnetic Material

Name:　　　　　　　　Student ID No.:　　　　　　Score:

Date:　　　　　　　　　Partner:　　　　　　　　　Teacher's signature:

1. Purpose

2. The Data recording and processing

(1) Measurement of the fundamental magnetization curve (Table 2.19.1).

The vertical sensitivity of CH1: $D_x =$ 　　　　(volts/div).

The vertical sensitivity of CH2: $D_y =$ 　　　　(volts/div).

Table 2.19.1　The fundamental magnetization curve

x/Div	$H/(\text{A} \cdot \text{m}^{-1})$	y/Div	B/T

(2) Measurement of the saturation hysteresis loop (Table 2.19.2).

The vertical sensitivity of CH1: $D_x =$ (volts/div).

The vertical sensitivity of CH2: $D_y =$ (volts/div).

Table 2.19.2 The saturation hysteresis loop

x/Div	$H/(\mathrm{A \cdot m^{-1}})$	y/Div	B/T

(3) Relative parameters (Table 2.19.3).

Table 2.19.3 Recording of relative parameters

N	R_1	n	R_2	C

(4) Measure (using a vernier caliper) the dimensions of the rectangular iron core (Table 2.19.4).

Table 2.19.4 Measurement of the dimensions of the rectangular iron core (see Fig. 2.19.5)

a/cm	a'/cm	b/cm	b'/cm	h/cm	d/cm
	$l = a + a' + b + b' =$			$A = h \times d =$	

Note: l is the mean magnetic path length; A is the cross-sectional area of rectangular iron core.

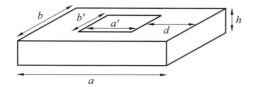

Fig. 2.19.5   The size of the rectangular iron core

(5) Plot the fundamental magnetization curve and the saturation hysteresis loop on graph paper in the same coordinate system.

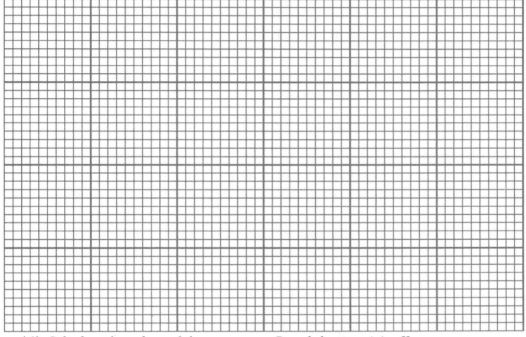

(6) Calculate the values of the remanence B_r and the coercivity H_c.

3. Discussion

2.20 Hall Effect and Its Application in Measuring the Magnetic Field

Hall Effect is a process in which a transverse electric field is developed in a solid material when the material carrying an electric current is placed in a magnetic field that is perpendicular to the current. Hall Effect was discovered by Edwin Herbert Hall in 1879. This effect is the basis of many practical applications and devices such as magnetic field measurements, and position and motion detectors.

1. Purpose

(1) To master the basic principle of Hall effect, the definition of Hall sensitivity and the definition of Hall coefficient;

(2) To understand the existence of systematic errors when measuring Hall voltage, and to learn the corresponding elimination methods.

2. Principle

(1) Lorentz force and Hall effect.

In fact, Hall effect results from the Lorentz force acting on moving charges and deflecting their motion. The deflected charges may accumulate on both sides of a path of the conducting sample, along which the charges propagate through. Due to the accumulation of charges, the built-in static electric field becomes stronger and stronger until a dynamic equilibrium state is established eventually, which prevents the further accumulation of charges, as shown in Fig. 2.20.1.

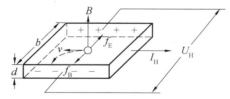

Fig. 2.20.1　The principle of Hall effect: (1) the accumulation of moving charges (say the electrons) in the magnetic field on both sides of a sample; (2) the establishment of dynamic equilibrium between Lorentz force and electric force; (3) the formation of Hall voltage

The Lorentz force F_L exerted by magnetic field B on a moving charge e can be expressed by a vector product:

$$F_L = ev \times B \qquad (2.20.1)$$

which is in quantity equal to $evB\sin\theta$, where v, B are the magnitudes of velocity and magnetic field respectively; θ is angle between the velocity vector and the vector of magnetic field, namely v and B. The orientation of the vector product is perpendicular to the plane determined by the two vectors. If the two vectors are mutually orthogonal, the product reaches the maximum evB.

In equilibrium the Lorentz force acting on a charge is offset by the electric force resulting from built-in field, which means

$$F_H = F_L \text{ or } eE_H = ev \times B \qquad (2.20.2)$$

When v and B are orthogonal, we have $E_H = vB$.

One can evaluate the current, which is actually called the Hall operating current I_H, formed by the directional movement of charges with concentration n passing through a cross section (the $d \times b$ in Fig. 2.20.1) with an average velocity $\bar{v}$:

$$I_H = ne\bar{v}bd \qquad (2.20.3)$$

Thus the Hall voltage U_H is easier to be obtained from above relations:

$$U_H = E_H b = \frac{1}{ned} I_H B = K_H I_H B \qquad (2.20.4)$$

where K_H is the Hall sensitivity, determined by

$$K_H = \frac{U_H}{I_H B} = \frac{1}{ned} \qquad (2.20.5)$$

From above, the Hall voltage is proportional to the product of Hall operating current I_H and strength of magnetic field B. The coefficient K_H has a unit of mV/(mA · T). For convenience, usually we definite a Hall coefficient R_H, which is derived from the Hall sensitivity:

$$R_H = K_H d = \frac{1}{ne} \qquad (2.20.6)$$

In fact, R_H is linked only to the material parameters rather than the structure parameters such as d.

From the expression of U_H, one can clearly see why in the practice the thin samples with lower concentration are preferentially employed, because the smaller values of n and d will result in higher U_H.

It should be pointed out that the Hall voltage is different from the voltage drop when a current flows in the resistant conductor. The Hall voltage is acquired in the transverse direction of the current. From the Hall sensitivity K_H or Hall coefficient R_H, one can get the information about the conducting type of the semiconductor chips.

(2) The systematic errors in measuring the Hall voltage and their elimination.

①The effect of nonsymmetrical electrodes. Arising from the fabrication technique the two welded electrodes to measure the Hall voltage usually are not located in the same plane which has the identical potential (the isopotential plane) when the current I_H flows through the Hall sample, see the Fig. 2.20.2, which may result in additional voltage drop $U_0 = I_H r$, where r denotes a resistance between the two isopotential planes. One can use the so-called symmetric method of measurement to eliminate such a kind of systematic error since U_0 is only dependent on the direction of the current I_H but not the magnetic field B. Changing the sign of U_0 by changing the direction of current I_H one may obtain the two outputs of voltage between the two electrodes. By averaging the two voltages the effect can be eliminated because in the first measurement the output may be $U_H - U_0$ while in the second measurement it becomes $U_H + U_0$.

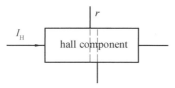

Fig. 2.20.2   The electrodes on the different isopotential planes of Hall component

②The Ettingshausen's effect. From the microscopic view, the speed of any carrier is not identical and just equal to the average speed $\bar{v}$ which decides the strength of Hall field. Only for the carrier with average speed $\bar{v}$ can be in balance. The carrier with a speed lower than $\bar{v}$ may deflect oppositely to that of the carrier with a speed larger than $\bar{v}$, which introduces a temperature difference electromotive force U_E, proportional to the product of current I_H and magnetic field B, namely $U_E \propto I_H B$. Changing the direction of I_H or B, one can change the direction of U_E so the symmetric method of measurement is invalid to eliminate the effect.

③Righi-Leduc's effect. If there is a temperature gradient along direction of I_H, it may induce a diffusion current I_d in this direction and results in an additional electromotive force $U_{RL} \propto I_d B$, which is related to the direction of B but I_H.

④Nernst's effect. The above-mentioned diffusion current may directly result in an additional voltage U_N, $U_N \propto I_d B$ whose direction is independent on I_H.

In the practice considering all effects the Hall voltage U_H must be measured via the average of absolute values for four voltages, U_1, U_2, U_3, U_4 which is obtained from all the combination of directions of I_H and B, namely $U_1 = U_{+B}^{+I_H}, U_2 = U_{-B}^{+I_H}, U_3 = U_{-B}^{-I_H}, U_4 = U_{+B}^{-I_H}$. Here the sigh of "+" and "−" means the different direction, so we have

$$U_H = \frac{|U_1| + |U_2| + |U_3| + |U_4|}{4} = \frac{|U_{+B}^{+I_H}| + |U_{-B}^{+I_H}| + |U_{-B}^{-I_H}| + |U_{+B}^{-I_H}|}{4} \quad (2.20.7)$$

3. Instruments

The experiment setup for study of the Hall effect is shown in Fig. 2.20.3. The Hall voltage U_H can be measured by adjusting the value of I_H (the working current flows through the Hall sample) and I_B (excitation current linked to the coil on the electromagnet, in the gap of which the variable magnetic field B is provided). Fig. 2.20.4 shows the circuit principle, in which the two steering switches are used to change the direction of working current I_H and magnetic field B by changing direction of I_B.

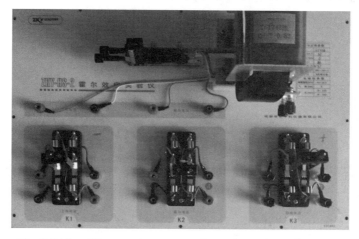

Fig. 2.20.3　The experiment setup for study on the Hall effect: front controlpanel and board for connecting devices with directional switches

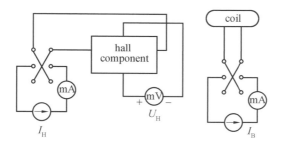

Fig. 2.20.4　The circuit principle for study on the Hall effect

4. Experiment contents

(1) Connect the circuit.

(2) Choose I_B = 400 mA (B = 100 mT), measure U_H by changing I_H, then complete Table 2.20.1 in the experimental report.

(3) Choose I_H = 3.0 mA, measure U_H by changing I_B, then complete Table 2.20.2 in the experimental report.

(4) Plot $U_H - I_H$, calculate the slope of the linear fitting line K_1 with graphic method, then calculate the Hall sensitivity $K_H = K_1/B$.

(5) Plot $U_H - B$, calculate the slope of the linear fitting line K_2 with graphic method, then calculate the Hall sensitivity $K_H = K_2/I_H$.

5. Discussion

How to judge the conduction type of Hall chip in this experiment according to the direction of B, I_H and U_H, and draw a picture to explain it.

Part 2   Topics of Experiments

Experimental Report

Hall Effect and Its Application in Measuring Magnetic Field

Name: Student ID No.:   Score:

Date: Partner: Teacher's signature:

1. Purpose

2. Experimental data and processing

(1) Measure the Hall voltage U_H by increasing I_H (Table 2.20.1).

Table 2.20.1   Measure the Hall voltage U_H by increasing I_H, keep $I_B = 400$ mA, $B = 100.0$ mT

I_H/mA	U_H/mV				
	$U_1 = U_{+B}^{+I_H}$	$U_2 = U_{-B}^{+I_H}$	$U_3 = U_{-B}^{-I_H}$	$U_4 = U_{+B}^{-I_H}$	$\bar{U}_H$
2.0					
2.5					
3.0					
3.5					
4.0					

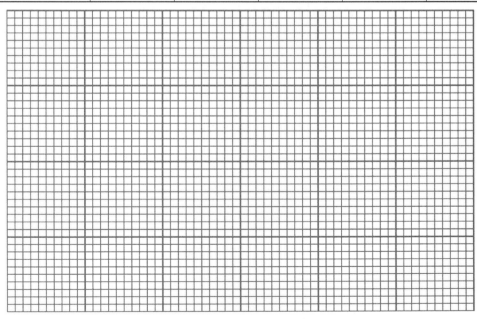

①Plot the experimental data in a graph (U_H versus I_H).

②Calculate the slope of the linear fitting line K_1 with graphic method.

③Calculate Hall sensitivity $K_H = K_1/B$. Don't forget the unit.

(2) Measure the Hall voltage U_H by increasing I_B (Table 2.20.2).

Table 2.20.2 Measure the Hall voltage U_H by increasing I_B, keep I_H = 3.0 mA

I_B/mA	B/mT	U_H/mV		
		$U_1 = U_{+B}^{+I_H}$	$U_4 = U_{+B}^{-I_H}$	$\bar{U}_H$
100	25.0			
200	50.0			
300	75.0			
400	100.0			
500	125.0			

①Plot the experimental data in a graph (U_H versus B).

②Calculate the slope of the linear fitting line K_2 with graphic method.

③Calculate Hall sensitivity $K_H = K_2/I_H$. Don't forget the unit.

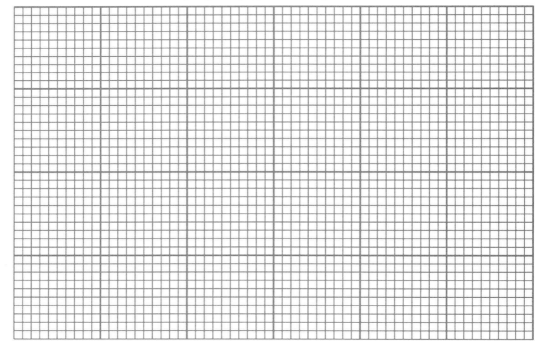

3. Discussion

2.21 Measuring the Focal Length of Thin Lens

1. Purpose

(1) To study the image-forming properties of a lens;
(2) To determine the focal length of a single thin lens.

2. Principle

(1) The focal length of a lens.

A thin lens may be defined as one whose thickness is considered small in comparison with the distances generally associated with its optical properties. Such distances are, for example, radii of curvature of the two spherical surfaces, primary and secondary focal lengths, and object and image distances.

There are two kinds of thin lenses, convex (converging) and concave (diverging) lens. A convex lens is one which is thicker at the center than at the periphery and converges incident parallel rays to a real focus on the opposite side of the lens as shown in Fig. 2.21.1(a). Convex lenses have positive focal lengths. Common shapes of convex lenses are: ①biconvex; ②convex − concave; ③plano-convex. A concave lens is thinner at the center than at the periphery and diverges the light from a virtual focus on the same side of the lens as shown in Fig. 2.21.1(b). Concave lenses have negative focal lengths and are thickest at the edges. Common shapes of diverging lenses are: ①biconcave; ②convex-concave; ③plano-concave.

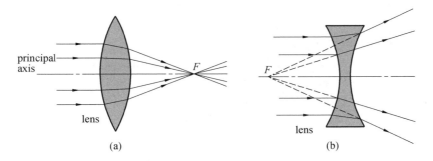

Fig. 2.21.1 Principal focus of converging lens and diverging lens

The principal axis of a lens is a line drawn through the center of the lens perpendicular to the face of the lens. The principal focus is a point on the principal axis through which incident rays parallel to the principal axis pass, or appear to pass, after refraction by the lens.

The focal length of a lens is the distance from the optical center of the lens to the principal focus. If two thin lenses with focal lengths of f_1 and f_2, respectively, are placed in contact to be used as a single lens, the focal length F of the combination is given by the equation

$$\frac{1}{F} = \frac{1}{f_1} + \frac{1}{f_2} \qquad (2.21.1)$$

(2) The Lensmaker's equation.

A converging lens may have the index of refraction n and two surface with radii of curvatures R_1 and R_2. The lens' thickness is small compared with R_1 and R_2 usually. The Lensmaker's equation, which links the focal length, the index of refraction, and the surface curvatures, via

$$\frac{1}{f} = (n - 1)\left(\frac{1}{R_1} - \frac{1}{R_2}\right) \qquad (2.21.2)$$

As the light travel from left to right through a lens, all convex surfaces are taken as having positive radius and all concave surfaces having negative radius. For an equiconvex lens like the one in Fig. 2.21.1(a), R_1 for the first surface is positive and R_2 for the second surface is negative, and therefore the focal length is positive.

(3) Image formation and the thin-lens equation.

When an object is placed on one side of a lens, an image is formed. We can determine the image position of an object using the famous Gaussian Lens Formula with

$$\frac{1}{f} = \frac{1}{u} + \frac{1}{v} \qquad (2.21.3)$$

where f, u, v denote the focal length, the object distance and the image distance, respectively. All quantities are measured to or from the center of the lens.

This equation can be applied to all possible combinations of real or virtual images formed by converging or diverging lenses with only a modification of plus or minus signs. The value of v should be negative when the image is a virtual image. Positive and negative signs are assigned for focal lengths of a converging lens and a diverging lens, respectively. For a distant object, u may be considered to be infinity, and Eq.(2.21.3) gives $v = f$.

In Table 2.21.1 and Table 2.21.2, the characteristics of image of real objects formed by a convex or concave lens is summarized.

Table 2.21.1 Images of real objects formed by convex lens

Object	Image			
Location	Type	Location	Orientation	Relative Size
$\infty > u > 2f$	Real	$f < v < 2f$	Inverted	Minified
$u = 2f$	Real	$v = 2f$	Inverted	Same size
$f < u < 2f$	Real	$\infty > v > 2f$	Inverted	Magnified
$u = f$	$\pm \infty$	—	—	—
$u < f$	Virtual	$|v| > u$	Erect	Magnified

Part 2 Topics of Experiments

Table 2.21.2 Images of real objects formed by concave lens

Object	Image			
Location	Type	Location	Orientation	Relative Size
Anywhere	Virtual	$|v|<f, |v|<u$	Erect	Minified

(4) Lens aberrations.

Aberration is a deviation from the ideal focusing of light by a lens. As a result, the image is not sharply focused. Some of the problems may lie with the geometry of the lens in use. Other aberrations may occur because different wavelengths of light are not refracted at the same position. The latter problem is called dispersion of light. This is clearly visible when white light passes through a prism.

①Spherical aberration. Spherical aberration appears because the light rays that are near and far from the principal axis focus at different focal points. Fig. 2.21.2 illustrates the spherical aberration for parallel rays passing through a convex lens. Rays near the middle of the lens are imaged farther from the lens than rays near the edges. Spherical aberration can be minimized by using an aperture to reduce the effective area of the lens, so that only light rays near the axis are transmitted. This happens in most cameras equipped with an adjustable aperture to control the light intensity, and when possible, reduce spherical aberration. To compensate for the light loss of a smaller aperture, a longer exposure time is needed. Another way to minimize the effect of aberration is by combinations of a convex and concave lenses. The aberration of one lens can be compensated by the optical properties of the other lens.

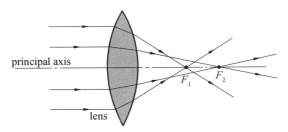

Fig. 2.21.2 Spherical lens aberration

②Chromatic aberration. In fact, the index of refraction of a material varies slightly with different wavelengths (dispersion); therefore, different wavelengths of light refracted by a lens focus at different points, giving rise to chromatic aberration. When white light passes through a lens, the transmitted rays of different wavelengths (colors) do not refract at the same angle. As a result, there is no common focal point for all the wavelengths. Fig. 2.21.3 shows that the images of different colors are produced at different locations on the principal axis. The blue color focuses before the red color. The chromatic aberration for a diverging lens is the opposite of that for a converging lens. Chromatic aberration can be greatly reduced by the use of a combination of convex and concave lenses made from two different types of glass with differing indexes of refraction. The lenses are chosen so that the dispersion produced by one lens is compensated by opposite dispersion produced by the other lens.

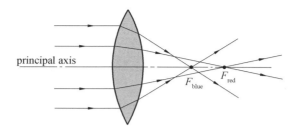

Fig. 2.21.3   Chromatic lens aberration

(5) Methods for measuring the focal length of a lens.

Besides the famous Gaussian Lens Formula as shown in Eq.(2.21.3), which can be used to determine the focal length of a convex lens by measuring the positions of the object and image, there are some other methods to measure the focal length of both convex and concave lenses.

①Measuring the focal length of a concave lens by forming a real image of a virtual object. There is a problem if we need to determine the focal length of a concave lens according to Eq.(2.21.3), since a diverging lens can only give a virtual image for a real object and such image can't be projected onto a screen so the position of the virtual image can't be determined. However, it is possible for a concave lens to produce a real image for a virtual object. All we need is an auxiliary convex lens whose focal length is available. This method for measuring the focal length of a concave lens is shown in Fig. 2.21.4.

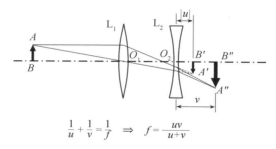

$$\frac{1}{u} + \frac{1}{v} = \frac{1}{f} \Rightarrow f = \frac{uv}{u+v}$$

Fig. 2.21.4   Principle of measuring the focal length of a concave
lens by using of a virtual object

②Measuring the focal length of a convex lens by the conjugate foci method. Examination of Fig. 2.21.5(a) shows that an enlarged image P is formed beyond the principal focus. If the object were placed at the position of the large image, a reversal light rays would give a diminished image O. These two interchangeable positions of object and image are called conjugate foci. The same result can be accomplished by moving the lens from position (Fig. 2.21.5(a)) to position (Fig. 2.21.5(b)), and leaving the position of the object and screen unchanged. In Fig. 2.21.5(b), a minified image P' is formed on the screen. Hence, the convex lens can be located at two positions l_1 and l_2 to produce an amplified and minified real image on the screen.

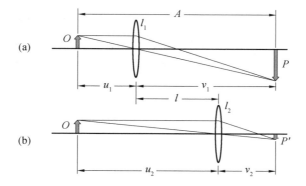

Fig. 2.21.5　Principle of measuring the focal length of a convex lens by the conjugate foci method

If A represents the fixed distance between the screen and the object, and l is the distance between the two possible positions of the lens, it is easily seen from Fig. 2.21.5(a) and Fig. 2.21.5(b) that

$$u_1 = \frac{A}{2} - \frac{l}{2} = \frac{A - l}{2} \tag{2.21.4}$$

$$v_1 = \frac{A}{2} + \frac{l}{2} = \frac{A + l}{2} \tag{2.21.5}$$

By substituting these values of u and v in Fig. 2.21.3, we get

$$f = \frac{A^2 - l^2}{4A} \tag{2.21.6}$$

This method of finding the focal length is especially valuable when using thick lenses, since the position of the optical center need not be known.

③Measuring the focal length of lens by self-collimation method. Place a plane mirror behind the convex lens perpendicular to the principal axis as shown in Fig. 2.21.6(a). Illuminate the object AB on the screen and adjust the position of the lens until a distinct image $A'B'$ appears on the screen. This requires that the rays be parallel on the opposite side of the lens from the object. The light rays emitting from the object AB are collimated before the mirror and reflected to form an inverted but identical real image $A'B'$ on the screen. The distance between the lens and the screen is just the focal length of the convex lens.

The situation is a little more complicated for a concave lens. As shown in Fig. 2.21.6(b), we need an additional convex lens L_1 to produce a real image $A'B'$, which acts as a virtual object for the concave lens L_2. The concave lens collimates the light before the mirror and the light is reflected to form an inverted but identical real image $A''B''$ on the screen. The distance between the concave lens and the real image $A'B'$ is the focal length of the concave lens, namely $|f'| = O'B'$.

Physics Laboratory Manual

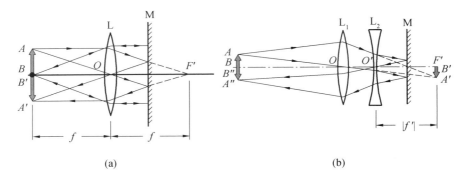

Fig. 2.21.6 Principle of measuring the focal length of a lens by the self-collimation method

3. Apparatus

Optical bench with meter stick, a convex(converging) lens, a concave (diverging) lens, lens mounts, a plane mirror, electric lamp, object screen, image screen.

4. Procedures

(1) To measure the focal length of the convex lens.

①The self-collimation method. Fix the electric lamp and the object screen at one end of the optical bench. Mount the convex lens some distance away from the object screen and the plane mirror 5 cm away from the lens. Turn on the lamp to illuminate the object. Record the position of the object screen B. Adjust the position of the lens until a distinct image of the object appears on the object screen. Make five independent measurements, record the positions of the lens O. Calculate the focal length.

②The conjugate foci method. Use the same lens here as was used in Step 1. Place the image screen at some convenient distance toward the other end so that the distance between them is more than $4f$. Record the positions of the object O and image screen P. Adjust the lens at the two positions for image formation, and make sure both the amplified and the diminished images are clear and distinct. Record the two positions of the lens. Make five independent measurements and calculate the focal length from Fig. 2.21.6.

(2) To measure the focal length of the concave lens.

①Focal length measured by virtual object and real image. Use the convex lens to form a distinct real image on the screen. Make five measurements and record the positions of the screen B'. Place the concave lens between the convex lens and the image screen. Adjust the position of the screen until the new image is distinct. Make five independent measurements and record the position of the screen B''. Calculate the focal length from Fig. 2.21.4.

②Focal length measured by the self-collimation method. Use the convex lens to form a distinct real image on the screen. Make five measurements and record the positions of the screen B'. Remove the image screen and mount the concave lens and the plane mirror according to Fig. 2.21.6(b). Adjust the position of the concave lens until a distinct image of the object shows on the object screen. Make five independent measurements, record the positions of the lens O', and calculate the focal length.

5. Discussion

(1) In real case the size of image may influence the accuracy for reading the position. For example, in Fig. 2.21.4, if the real image formed by the convex lens is larger than the object, what phenomenon would you see when observing $A''B''$? Is there any influence on determining the focal length of the concave lens?

(2) Is there any limitation for position of the convex lens L_1 in measuring the focal length of a concave lens by using the self-collimation method (Fig. 2.21.6(b))?

Physics Laboratory Manual

Experimental Report

Measuring the Focal Length of Thin Lens

Name: Student ID No.: Score:

Date: Partner: Teacher's signature:

1. Purpose

2. Data record and analysis

(1) Measure the focal length of convex lens by self-collimation method as Fig. 2.21.6(a) and fulfill Table 2.21.3.

Table 2.21.3 Measure the focal length of convex lens by self-collimation method (in cm)

Trial	1	2	3	4	5	Average
B						
O						

Calculations (show work).

(2) Measure the focal length of convex lens by conjugate foci method as Fig. 2.21.5 and fulfill Table 2.21.4.

Table 2.21.4 Measure the focal length of convex lens by conjugate foci method (in cm)

Trial	1	2	3	4	5	Average
O						
P						
l_1						
l_2						

Calculations (show work).

(3) Measure the focal length of concave lens by virtual object and real image as Fig. 2.21.4 and fulfill Table 2.21.5.

Table 2.21.5 Measure the focal length of concave lens by virtual object and real image (in cm)

Trial	1	2	3	4	5	Average
B'						
O_2						
B''						

Calculations (show work).

(4) Measure the focal length of concave lens by self-collimation method as Fig. 2.21.6(b) and fulfill Table 2.21.6.

Table 2.21.6 Measure the focal length of concave lens by self-collimation method (in cm)

Trial	1	2	3	4	5	Average
B'						
O'						

Calculations (show work).

3. Discussion

2.22 Self-assembling a Microscope and a Telescope

1. Purpose

(1) To master the operational principle of two simplest optical instruments — microscope and telescope;

(2) To master the methods to measure focal lengths of thin lenses such as convex and concave lenses by using of a telescope focusing on infinity.

2. Principle

Microscope and telescope are the most fundamental and frequently-used optical instruments, they are also the ones with the simplest structure. So it is necessary to master their structures and operational principle. This is benefit for further understanding more complex optical instruments.

(1) The Magnifier.

The magnifier is a positive (convex) lens that can increase the size of the retinal image compared to that is formed with the unaided eye. The apparent size of any object as seen with the unaided eye depends on the angle subtended by the object on the retina. As the object is brought closer to the eye, from A to B to C in the Fig. 2.22.1, the retinal image becomes larger and larger.

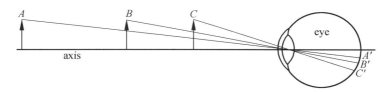

Fig. 2.22.1   Size of the retinal image depends on the distance of an object from the eye

There is a limit to how close an object may come to the eye if the latter is still to have sufficient accommodation to produce a sharp image. Although the nearest point varies widely with various individuals, 25.0 cm is taken to be the standard <u>near point</u>, sometimes called the <u>distance of most distinct vision</u> for an average level. At this distance, as indicated in Fig. 2.22.2(a), the angle subtended by object or image will be called θ. If a convex lens is now placed before the eye as shown in diagram Fig. 2.22.2(b), the object y can be brought much closer to the eye and an image subtending a larger angle θ' will be formed on the retina. What the convex lens has done is to form a virtual image y' of the object y and the eye is able to focus upon this virtual image. Any lens used in this manner is called a <u>magnifier</u> or <u>simple microscope</u>. If the object y is located at F, the focal point of the magnifier, the virtual image

y' will be located at infinity and the eye will be accommodated for distant vision as is illustrated in Fig. 2.22.2(c). If the object is properly located a short distance inside of F as in diagram Fig. 2.22.2(b), the virtual image may be formed at the distance of most distinct vision, and a slightly greater magnification is obtained.

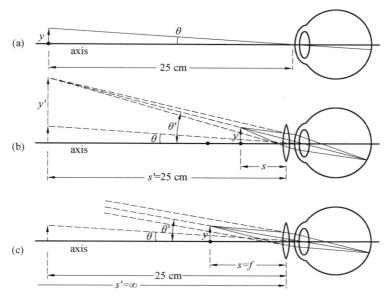

Fig. 2.22.2 The distance of most distinct vision and the image with a magnifier

The angular magnification (magnifying power) M is defined as the ratio of the angle θ' subtended by the image to the angle θ subtended by the object, then

$$M = \frac{\theta'}{\theta} \qquad (2.22.1)$$

From diagram Fig. 2.22.2(b), the object distance s is obtained by the regular thin-lens formula as

$$\frac{1}{s} + \frac{1}{-25} = \frac{1}{f} \quad \text{or} \quad \frac{1}{s} = \frac{25+f}{25f} \qquad (2.22.2)$$

From the right triangles, the angle θ and θ' are given by

$$\tan \theta = \frac{y}{25}, \quad \tan \theta' = y \frac{25+f}{25f} \qquad (2.22.3)$$

For small angles, the tangents can by replaced by the angles themselves to give approximate relations:

$$\theta = \frac{y}{25} \quad \text{and} \quad \theta' = y \frac{25+f}{25f} \qquad (2.22.4)$$

The magnification is

$$M = \frac{\theta'}{\theta} = \frac{25}{f} + 1 \qquad (2.22.5)$$

In diagram Fig. 2.22.2(c), the object distance s is equal to the focal length, and the small angles are given by

$$\theta = \frac{y}{25} \text{ and } \theta' = \frac{y}{f} \qquad (2.22.6)$$

And the magnification is

$$M = \frac{\theta'}{\theta} = \frac{25}{f} \qquad (2.22.7)$$

The angular magnification is therefore larger if the image is formed at the distance of most distinct vision. For example, let the focal length of a magnifier be 1 inch (2.5 cm). For these two extreme cases, Eq.(2.22.6) and Eq.(2.22.7) give two magnifications

$$M = \frac{25}{2.5} + 1 = 11 \times, \ M = \frac{25}{2.5} = 10 \times \qquad (2.22.8)$$

Because magnifiers usually have short focal lengths and therefore give approximately the same magnifying power for object distances between 25.0 cm and infinity, the simpler expression $25/f$ is commonly used in labeling the power of magnifiers. Hence a magnifier with a focal length of 2.5 cm will be marked 10 × and another with a focal length of 5.0 cm will be marked 5 ×, etc.

It may seem that we can make the angular magnification as large as we like by decreasing the focal length f. In fact, the aberrations of a simple double-convex lens set a limit to M of about 3 × to 4 ×. If these aberrations are corrected, the angular magnification may be made as great as 20 ×. When greater magnification than this is needed, we usually use a compound microscope.

CAUTION

Don't confuse the angular magnification M with the lateral magnification m. Angular magnification is the ratio of the angular size of an image to the angular size of the corresponding object; lateral magnification refers to the ratio of the height of an image to the height of the corresponding object. For the situation shown in Fig. 2.22.2(b), the angular magnification is about 3 × since the object subtends an angle about three times larger than that in Fig. 2.22.2(a). The lateral magnification in Fig. 2.22.2(c) is infinite because the virtual image is at infinity, but that doesn't mean that the object looks infinitely large through the magnifier! (That's why we didn't attempt to draw an infinitely large object image in Fig. 2.22.2(c).) When dealing with a magnifier, M is useful but m is not.

(2) The Microscope.

When we need greater magnification than we can get with a simple magnifier, the instrument that we usually use is the microscope, in fact which was invented by Galileo in 1610. In its simplest form, an optical microscope consists of two lenses, one of very short focal length called the objective lens, and the other with longer focus is call the ocular or eyepiece. Both lenses actually contain several elements to reduce aberrations. The essential elements of a microscope are shown in Fig. 2.22.3(a). To analyze this system, we use the principle that an image formed by one optical element such as a lens or mirror can serve as the object for a second element.

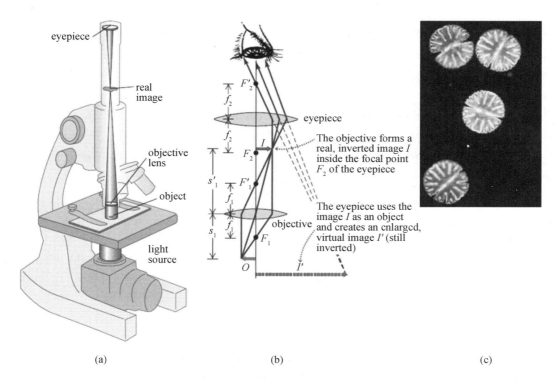

Fig. 2.22.3 (a) Elements of a microscope. (b) The object O is placed just outside the first focal point of the objective (the distance has been exaggerated for clarity). (c) This microscope image shows single-celled organisms about 2×10^{-4} m (0.2 mm) across. Typical light microscopes can resolve features as small as 2×10^{-7} m, comparable to the wavelength of light

The object O to be viewed is placed just beyond the first focal point of the objective, a converging lens that forms a real and enlarged image I (Fig. 2.22.3(b)). In a properly designed instrument, this image lies just inside the first focal point F_2 of a second converging lens — the eyepiece or ocular. The eyepiece acts as a simple magnifier, and forms a final virtual image I' of I. The position of I' may be anywhere between the near and far points of the eye. Both the objective and the eyepiece of an actual microscope are highly corrected compound lenses with several optical elements, but for simplicity we show them here as simple thin lenses. As for a simple magnifier, what matters when viewing through a microscope is the angular magnification M. The overall angular magnification of the compound microscope is the product of two factors. The first factor is the lateral magnification m_1 of the objective, which determines the linear size of the real image I; the second factor is the angular magnification M_2 of the eyepiece, which relates the angular size of the virtual image seen through the eyepiece to the angular size that the real image I would have if you viewed it without the eyepiece. The first of these factors is given by

$$m_1 = -\frac{s_1'}{s_1} \qquad (2.22.9)$$

where s_1 and s_1' are the object and image distances, respectively, for the objective lens. Ordinarily, the object is very close to the focal point, and the resulting image distance s_1' is very great in comparison to the focal length f_1 of the objective lens. Thus s_1 is approximately equal to f_1 and we can write $m_1 = -s_1'/f_1$. The real image I is close to the focal point F_2 of the eyepiece, so to find the eyepiece angular magnification, we can use $M_2 = 25/f_2$, where f_2 is the focal length of the eyepiece (considered as a simple lens). The overall angular magnification M of the compound microscope (apart from a negative sign, which is customarily ignored) is the product of the two magnifications:

$$M = m_1 M_2 = -\frac{25 \cdot s_1'}{f_1 f_2} \qquad (2.22.10)$$

where s_1', f_1 and f_2 are measured in centimeters. The final image is inverted with respect to the object. Because the focal length f_2 is much less than s'_1, one may often use the length of microscope tube L (the distance between ocular and objective) to replace s_1' in the magnification formula:

$$M = m_1 M_2 = -\frac{25 \cdot L}{f_1 f_2} \qquad (2.22.11)$$

Microscope manufacturers usually specify the values of m_1 and M_2 for microscope components rather than the focal lengths of the objective and eyepiece.

It is clear that the angular magnification of a microscope can be increased by using an objective of shorter focal length f_1, thereby increasing m_1 and the size of the real image I. Most optical microscopes have a rotating "turret" with three or more objectives of different focal lengths so that the same object can be viewed at different magnifications. The eyepiece should also have a short focal length f_2 to help to maximize the value of M.

To take a photograph using a microscope (called a photomicrograph or micrograph), the eyepiece is removed and a camera placed so that the real image I falls on the camera's CCD array or film. Fig. 2.22.3(c) shows such a photograph.

(3) Telescopes.

①Basic principle. The optical system of a refracting telescope is similar to that of a microscope. In both instruments the image formed by an objective is viewed through an eyepiece. The difference is that the telescope is used to view large objects at large distances, while the microscope is used to view small objects close at hand.

Historically the first telescope was probably constructed in Holland in 1608 by an obscure spectacle-lens grinder, Hans Lippershey. A few months later Galileo, upon hearing that objects at a distance could be made to appear close at hand by means of two lenses, designed and made with his own hands the first authentic telescope. The elements of this telescope are still in existence and may be seen on exhibit in Florence.

②Astronomical telescope. An astronomical telescope is shown in Fig. 2.22.4. The

Part 2 Topics of Experiments

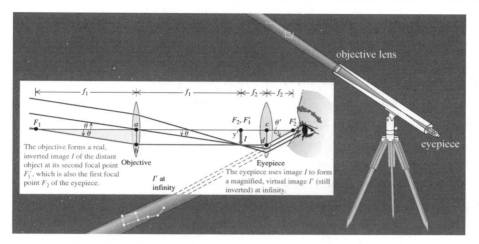

Fig. 2.22.4 The structure of an astronomical (refracting) telescope

objective forms a reduced real image I of the object, and the eyepiece forms an enlarged virtual image of I. As with the microscope, the image may be formed anywhere between the near and far points of the eye. Objects viewed with a telescope are usually so far away from the instrument that the first image I is formed very nearly at the second focal point of the objective. This image is the object for the eyepiece. If the final image formed by the eyepiece is at infinity (for the most comfortable viewing by a normal eye), the first image must be at the first focal point of the eyepiece. The distance between the objective and the eyepiece is the length of the telescope tube, which is therefore the sum of the focal lengths of the objective and the eyepiece. The angular magnification M of a telescope is defined as the ratio of the angle subtended at the eye by the final image to the angle subtended at the (unaided) eye by the object. We can express this ratio in terms of the focal lengths of the objective and the eyepiece. In Fig. 2.22.4, the ray passing through F_1 the first focal point of the objective, and through the second focal point of the eyepiece, has been emphasized. The object subtends an angle at the objective and would subtend essentially the same angle at the unaided eye. Also, since the observer's eye is placed just to the right of the focal point the angle subtended at the eye by the final image is very nearly equal to the angle. Because bd is parallel to the optic axis, the distances ab and cd are equal to each other and also to the height of the image I. Because θ and θ' are small, they may be approximated by their tangents. From the right triangles $F_1 ab$ and $F'_2 cd$ we obtain

$$\theta = \frac{-y'}{f_1}, \quad \theta' = \frac{y'}{f_2} \qquad (2.22.12)$$

and the angular magnification M is

$$M = \frac{\theta'}{\theta} = \frac{-y'/f_2}{y/f_1} = -\frac{f_1}{f_2} \qquad (2.22.13)$$

The angular magnification M of a telescope is equal to $-f_1/f_2$ — the ratio of the focal length of the objective to that of the eyepiece. The negative sign shows that the final image is inverted.

An inverted image is no particular disadvantage for astronomical observations. When we use a telescope or binoculars on earth, though, we want the image to be right side up. Inversion of the image is accomplished in prism binoculars by a pair of 45°-45°-90° totally reflecting prisms called Porro prisms. Porro prisms are inserted between objective and eyepiece, as shown in Fig. 2.22.5. The image is inverted by the four internal reflections from the faces of the prisms at 45°. The prisms also have the effect of folding the optical path and making the instrument shorter and more compact than it would otherwise be. Binoculars are usually described by two numbers separated by a multiplication sign, such as 7 × 50. The first number is the angular magnification M, and the second is the diameter of the objective lenses (in millimeters). The diameter determines the light-gathering capacity of the objective lenses and thus the brightness of the image.

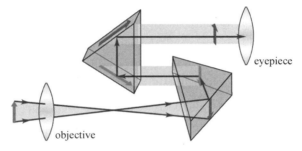

Fig. 2.22.5   Inversion of an image in prism binoculars

③Reflecting Telescopes. In the reflecting telescope (Fig. 2.22.6), the objective lens is replaced by a concave mirror. In large telescopes, this scheme has many advantages, both theoretical and practical. Mirrors are inherently free of chromatic aberrations (the dependence of focal length on wavelength), and spherical aberrations (associated with the paraxial approximation) are easier to correct than for a lens. The reflecting surface is sometimes parabolic rather than spherical. The material of the mirror need not be transparent, and the mirror can be made more rigid than a lens, which can be supported only at its edges.

Because the image in a reflecting telescope is formed in a region traversed by incoming rays, it can be observed directly with an eyepiece only by blocking off part of the incoming beam (Fig. 2.22.6(a)), an arrangement that is practical only for the very large telescopes. Alternative schemes use a mirror to reflect the image through a hole in the mirror or out the side, as shown in Fig. 2.22.6(b) and Fig. 2.22.6(c). This scheme is also used in some long-focal-length telephoto lenses for cameras. In the context of photography, such a system is called a catadioptric lens — an optical system containing both reflecting and refracting elements. The Hubble Space Telescope(Fig. 2.22.6(d)) has a mirror 2.4 m in diameter, with an optical system similar to that in Fig. 2.22.6(b). It is very difficult and costly to fabricate large mirrors and lenses with the accuracy needed for astronomical telescopes. Lenses larger than 1 m in diameter are usually not practical. A few current telescopes have single mirrors

Part 2 Topics of Experiments

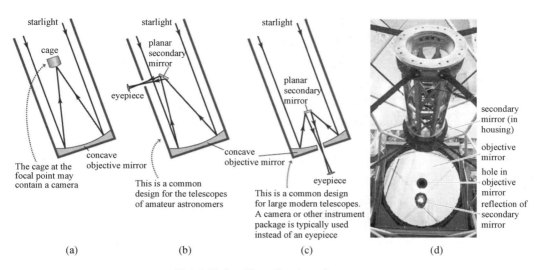

Fig. 2.22.6 The reflecting telescope

over 8 m in diameter. The largest single-mirror reflecting telescope in the world, the Large Binocular Telescope on Mt Graham in Arizona, will have two 8.4 m mirrors on a common mount, the first mirror went into operation in October 2005. Another technology for creating very large telescope mirrors is to construct a composite mirror of individual hexagonal segments and use computer control for precise alignment. The twin Keck telescopes, atop Mauna Kea in Hawaii, have mirrors with an overall diameter of 10 m, made up of 36 separate 1.8 m hexagonal mirrors. The orientation of each mirror is controlled by computer to within one millionth of an inch. In the Southern Hemisphere, the Southern African Large Telescope (SALT) has an 11 m mirror composed of 91 hexagonal segments. It went into operation in September 2005.

④The role of the crosshair. Inside the tube of some optical instrument one may often find a crosshair close to the eyepiece, which in fact plays important role in adjustment and observation of the object. In the first step of adjustment, one should regulate the relation of the eyepiece and the crosshair in order clearly to sight the virtual image at the distance of most distinct vision. In the second step, one needs to regulate the relation between the objective and the object in order to let the image formed by objective just falls into the plane at which the crosshair is located (we call the status of will-adjustment the state of eliminating parallax). The crosshair plays a role of reference for observation or measurement.

Strictly speaking, the position of plane at which the image of the crosshair is located has a narrow range for personal reason from psychology and physiology. Without assistance of the crosshair, however, the final imaging quality of an optical system cannot be guaranteed.

3. Experiment contents

(1) Assemble a simple compound microscope with sufficiently high magnification ($M > 10$) using given optical elements.

Tips: the key point of experiment is to choose a correct combination of elements and change the magnification with fixed elements.

It is recommended that all the data (the position of the elements) are recorded in the coordinates of the metal ruler on the optical bench and filled in a blank (please make it by yourself). This recording method is also suggested in the following experiment.

(2) Assemble an astronomical telescope of Kepler's type (with two positive lenses) using given optical elements.

Tips: The Kepler's telescope is a confocal optical system with final inversed image. The tube length of telescope approximately equals to the algebraic sum of two focal lengths of the lenses from which the telescope is constructed.

(3) Using the well-tuned telescope measure the focal lengths of the elements unemployed in constructing the telescope.

Tips: the well-tuned telescope is such an optical system which is suitable for the incidence of parallel beam. The key point of experiment is how to form the parallel beam via an element to be measured.

It is not allowed to change the relative position of the elements in the telescope in the process of observation and measurement (Why?). The telescope may be simply illustrated as a black box in the schematic principle designed by yourself.

In Fig. 2.22.7, the method to measure the focal length of a positive lens is illustrated schematically via well-tuned telescope. While in Fig. 2.22.8, the focal length of a negative lens should be measured with assistance of a known positive lens by using "first converging then diverging" method. Please give the detailed procedure to calculate the focal length of the negative lens and clarify the experimental condition.

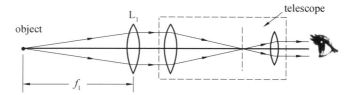

Fig. 2.22.7 The schematic principle to measure the focal length of a positive (converging) lens

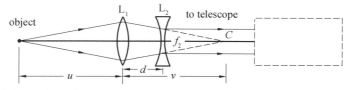

Fig. 2.22.8 The schematic principle to measure the focal length of a negative (diverging) lens with assistance of a known positive lens (the "first converging then diverging" method)

4. Discussion

(1) Is it possible to use the "first diverging then converging" method to generate the parallel beam for the well-tuned telescope focusing on infinity to measure the focal length of a negative lens with assistance of a known positive lens?

(2) Please give the detailed procedure to calculate the focal length of the negative lens. What kind of condition is needed to guarantee performing the measurement?

Experimental Report

Self-assembling a Microscope and a Telescope

Name: Student ID No.:   Score:

Date: Partner: Teacher's signature:

1. Purpose

2. The experimental data and processing

(1) Build the microscope.

Select the double-convex converging lenses (the ones with shorter-focus, why?) for the objective and eyepiece. Measure the focal lengths of eyepiece and objective f_e and f_o. Place the reticle at a convenient distance away from the eyepiece and make sure that the crosshairs can be seen clearly through the eyepiece (the eye is required as close as possible to the eyepiece). Place the objective at an adequate distance away from the crosshair plate (why?). Mount the object screen and adjust its position until a clear image can be seen through the eyepiece with no parallax on the reticle. Record the positions of the object, the lenses and the reticle. Calculate the object distance s_1 and image distance s_1'. Calculate the magnification M of the microscope and show your work.

f_e = _____ ;
Object distance s_1 = _____ ;
Image distance s_1' = _____ ;
f_o = _____ ;
Magnification M = _____ .
Calculations (show work).

(2) Build the astronomical telescope.

Use the combination of the eyepiece and the reticle used in step (1) as the ocular system. Use another double-convex lens with the longest focal length (why?) as the objective (it is suggested that its focal length is determined in advance by using the method to measure its object distance u and image distance v in the bench). Adjust the position of the objective until a clear image of a distant object is seen through the eyepiece. Calculate the angular magnification M.

  Object distance u = _____;
  Image distance v = _____;
  f_o = _____;
  Magnification M = _____.
  Calculations (show work).

(3) Measure the focal length of a convex lens.

Focus the telescope in step (2) at infinity. Mount the object and the convex lens with unknown focal length f_1 and adjust the position of L_1 (Fig. 2.22.7) until a distinct image is formed on the reticle without parallax. Record the positions of the object and L_1 and calculate its focal length.

  Positioa of object _____;
  Position of lens L_1_____;
  f_1 = _____.

(4) Measure the focal length of a diverging lens.

Use the telescope focused at infinity as you did in step (3). Mount the object and the converging lens L_1 to form a real image at C (Fig. 2.22.8). Place the diverging lens L_2 between the telescope and L_1. Adjust the position of L_2 until a distinct image is formed on the reticle without parallax. Record the positions of the object and L_2 and calculate its focal length f_2. Use the value of f_1 you get in step (3). Please pay much attention to the effect of relative positions between the object and the converging lens L_1.

Position of object _____ ;

Position of lens L_1 _____ ;

Position of image _____ ;

Position of L_2 _____ ;

$f_2 =$ _____ .

Calculations (show work).

3. Discussion

2.23 Measuring the Wavelengths of Light with a Transmission Diffraction Grating

The spectrometer is an instrument for precise measurement of deflection of light angle. A diffraction grating is an optical component with a periodic structure, which splits and diffracts light into several beams travelling in different directions. The directions of these beams depend on the spacing of the grating and the wavelength of the light, so that the grating acts as the dispersive element.

In this experiment, a grating spectrometer will be used for measuring diffraction angle and the wavelength of a sodium light source.

1. Purpose

(1) To learn the skill for adjustment of a grating spectrometer and master the function of all parts;

(2) To measure the wavelength of sodium light with a grating spectrometer.

2. Principle

In optics, a diffraction grating is an optical component with a periodic structure that splits and diffracts light into several beams travelling in different directions. The emerging coloration is a form of structural coloration. The directions of these beams depend on the spacing of the grating and the wavelength of the light so that the grating acts as the dispersive element. Because of this, gratings are commonly used in monochromators and spectrometers.

For practical applications, gratings generally have ridges or rulings on their surface rather than dark lines. Such gratings can be either transmissive or reflective. Gratings that modulate the phase rather than the amplitude of the incident light are also produced, frequently using holography.

Gratings may be of the "reflective" or "transmissive" type, analogous to a mirror or lens, respectively.

Diffraction can create "rainbow" colors when illuminated by a wide spectrum (e.g., continuous) light source.

An idealised grating is made up of a set of slits of spacing d, that must be wider than the wavelength of interest to cause diffraction. As shown in Fig. 2.23.1, assuming a plane wave of monochromatic light of wavelength λ with normal incidence (perpendicular to the grating), each slit in the grating acts as a quasi point-source from which light propagates in all directions (although this is typically limited to a hemisphere). After light interacts with the grating, the diffracted light is composed of the sum of interfering wave components emanating from each slit in the grating. At any given point in space through which diffracted light may pass, the path length to each slit in the grating varies. Since path length varies, generally, so do the phases of the waves at that point from each of the slits. Thus, they add or subtract from each other to create peaks and valleys through additive and destructive interference. When the path difference between the light from adjacent slits is equal to half the wavelength, $\lambda/2$, the waves are out of phase, and thus cancel each other to create points of minimum intensity. Similarly, when the path difference is λ, the phases add together and maxima or a complete constructive interference of the waves occurs. The K_{th} order maximum satisfies the following equation with

$$d\sin \partial_K = K\lambda, \quad K = 0, \pm 1, \pm 2, \pm 3, \cdots \quad (2.23.1)$$

where K is the order of the image maximum; d denotes the grating constant given by

$$d = \frac{1}{N} \quad (2.23.2)$$

with N being the number of lines per unit length.

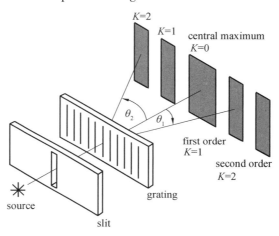

Fig. 2.23.1 Diffraction structure and pattern

Note that a grating has a "zero-th" order mode (where $K = 0$), in which there is no diffraction and a ray of light behaves according to the laws of reflection and refraction the same as with a mirror or lens, respectively.

The interference patterns are symmetric on either side of the central maximum. If the incident light is compound or complex light rather than monochromatic one say the white light, each order corresponds to a colored line or belt, all of which forms a spectrum, but "zero-th" order image is observed usually as a polychromatic light as the incident light, namely, no diffraction occurs.

3. Instruments

In this experiment, spectrometer, diffraction grating and a sodium discharge tube are needed.

In a spectrometer the following four basic parts matter: ① collimator and slit assembly;

② spectrometer table; ③ telescope; ④ scale, as shown in Fig. 2.23.2.

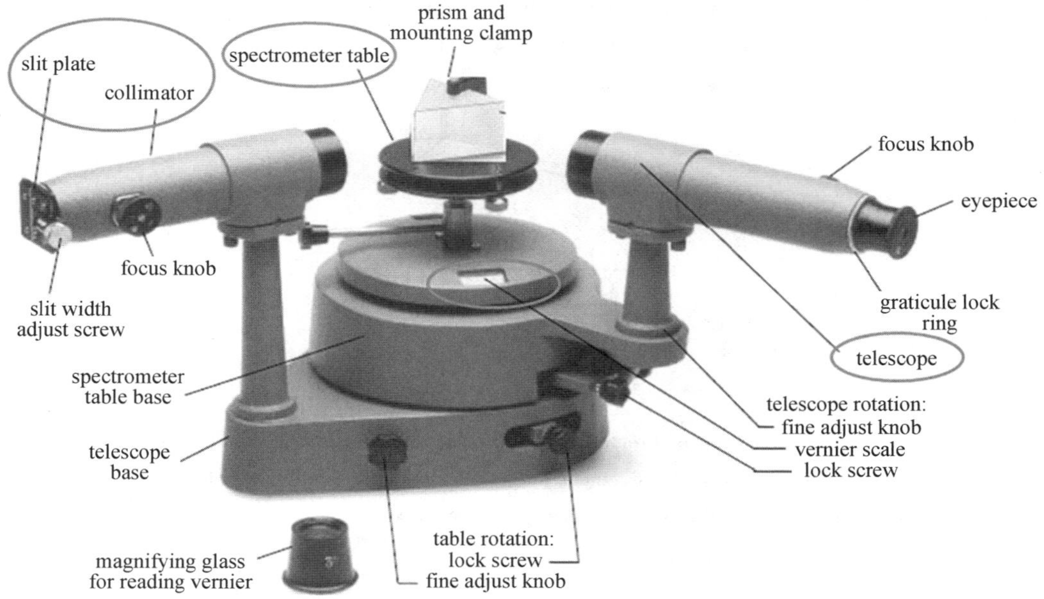

Fig. 2.23.2 The structure of a spectrometer

The telescope is a tube with the eyepiece at one end and a converging lens at the other, and crosshairs in the middle. The collimator is a tube with a slit of adjustable width at one end and a converging lens at the other. Light from a light source enters the collimator. The length of the collimator tube is made approaching the focal length of the lens so as to make the rays of the emerging light beam parallel.

In this experiment, sodium discharge tube is used to be a light source and the color of the incident light is yellow. Actually, there are two wavelengths in the sodium yellow light, namely 5 890 Å and 5 896 Å. Since the grating spread the beam into a spectrum, the two wavelengths can be observed in the higher order of diffraction.

4. Procedures

(1) Put the mirror on the spectrometer table, as shown in Fig. 2.23.3, adjust the spectrometer to meet these requirements (details see Part A):

①The telescope is adjusted suitable for parallel light rays and the optical axis of the telescope is perpendicular to the central axis of spectrometer strictly.

②The collimator emits the parallel light and the optical axis of the collimator is perpendicular to the central axis of spectrometer strictly.

(2) Put the grating on the spectrometer table instead of the mirror, as shown in Fig. 2.23.3, make the telescope opposite to the grating, adjust the reflected image coincide with the image of slit and crosshairs. Lock the screw.

(3) Find the diffraction pattern by rotating the telescope, as shown in Fig. 2.23.4, and

record the angular position for different orders ($K = -3, -2, -1, 0, 1, 2, 3$).

(4) Fill the Table 2.23.1 in the experimental report, and calculate the wavelengths.

(5) Observe the structure of "double-line" of sodium yellow light, calculate the corresponding wavelength and the wavelength difference, and compare it with $5\,896 - 5\,890 = 6$ (Å).

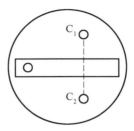

Fig. 2.23.3 The mirror (rectangle) on the spectrometer table (great circle) (one leveling screw is fixed, the straight line linking the other two leveling screws is perpendicular to the surface of the mirror)

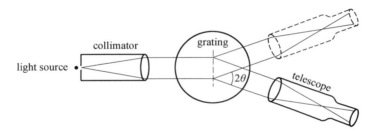

Fig. 2.23.4 Find the diffraction pattern by rotating the telescope

5. Part A. Adjustment of the spectrometer

(1) Visual coarse adjustment.

Use visual method for the coarse adjustment to make the optical axis of telescope and collimator perpendicular to the central axis of spectrometer approximately, and to make the table plane.

(2) Adjustment of the telescope, as shown in Fig. 2.23.5.

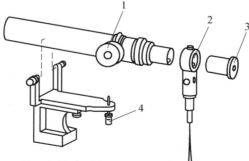

Fig. 2.23.5 The structure of a telescope

1—focus knob; 2—crosshairs; 3—eyepiece; 4—height – adjustment screw

①Adjust the eyepiece until the crosshairs are clear and sharp.

②Put the mirror on the spectrometer table, see Fig. 2.23.3, look through the telescope and turn the focus knob until the image of green cross is clear and sharp. That denotes the distance from lens to crosshairs is equal to the focal length of the lens. Now, the telescope is focused for parallel light rays.

③Turn the mirror 180°, make sure the two images of green light on both sides can be found. If one or none image can be viewed, do step of visual coarse adjustment again until two images are found.

④Repeat the following steps until the reflected images on both sides of the mirror coincide with the crosshair(point P), as shown in Fig. 2.23.6. Now, the optical axis of the telescope is perpendicular to the central axis of spectrometer strictly.

a. Correct half the difference using leveling screw C_1 (Fig. 2.23.3) of the spectrometer table and the other half using the height-adjustment screw of the telescope (Fig. 2.23.5).

b. Turn the telescope table 180°, so a green image reflected on the opposite side of the mirror can be found. Use half difference method again to make reflected image be coincident to point P. Correct half the difference using leveling screw C_2 (Fig. 2.23.3) of the table and the other half using the height-adjustment screw of the telescope (Fig. 2.23.5).

Fig. 2.23.6   The crosshairs of the telescope

(3) Adjustment of the collimator.

DO NOT change the height-adjustment screw and the focus screw of the telescope any case.

①Remove the mirror from the spectrometer table and turn on the light.

②Check that the collimator slit is partially open (use the slit width adjust screw).

③Make the telescope directly opposite the collimator.

④Look through the telescope, adjust the focus screw of the collimator and, if necessary, the rotation of the telescope until the slit image is sharp. That means the distance from slit to lens is equal to the focal length of the lens. Now, the collimator exits the parallel light.

⑤Use the height-adjustment screw of the collimator to make the image of slit in the middle of the visual field. Now the optical axis of the collimator is perpendicular to the central axis of spectrometer strictly.

6. Part B. How to read the angle

As shown in Fig. 2.23.7, the degree scale is 250°40′ (the smallest subdivision of the main scale is 20′); the vernier scale is 2′00″ (the smallest subdivision of the vernier scale is 30″).

$$\text{Total}: 250°40' + 2'00'' = 250°42'00''$$

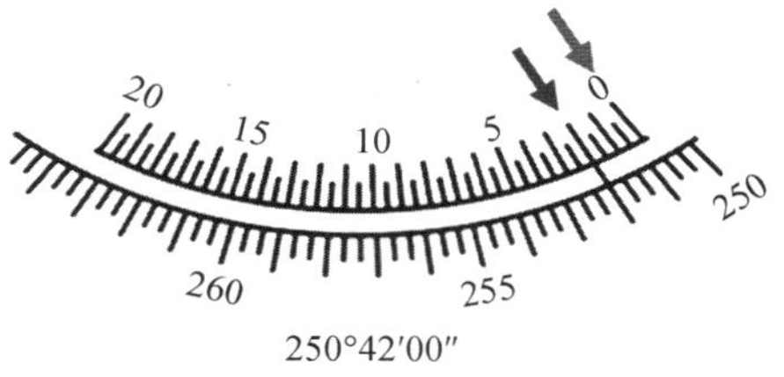

250°42′00″

Fig. 2.23.7   Vernier scale

7. Discussion

What requirements should the spectrometer meet before measurement?

Part 2　Topics of Experiments

Experimental Report

Measuring the Wavelengths of Light with a Transmission Diffraction Grating

Name:　　　　　　　　　Student ID No.:　　　　　　　Score:

Date:　　　　　　　　　　Partner:　　　　　　　　　　Teacher's signature:

1. Purpose

2. Experimental observing, data recording and processing

(1) Fill in the following Table 2.23.1 and perform the calculation.

Table 2.23.1　Measure a wavelength of sodium light with a grating spectrometer

K	Left window θ_{K_L}	Right window θ_{K_R}	$\theta_{1K} = \frac{1}{2}\|\theta_{+K\_L} - \theta_{-K\_L}\|$	$\theta_{2K} = \frac{1}{2}\|\theta_{+K\_R} - \theta_{-K\_R}\|$	$\bar{\theta}_K = \frac{1}{2}\|\theta_{1K} + \theta_{2K}\|$	$\lambda_K = \frac{d\sin\bar{\theta}_K}{K}$ /nm
+1						
−1						
+2						
−2						
+3						
−3						

Here, $d = 1/300$ mm. Calculate the average of wavelength and compare with a standard value $\lambda_0 = 589.3$ nm.

① $K = 1$, $\lambda_1 =$

② $K = 2$, $\lambda_2 =$

③ $K = 3$, $\lambda_3 =$

④ $\bar{\lambda} = \frac{1}{3}(\lambda_1 + \lambda_2 + \lambda_3) =$

⑤ $E = |\bar{\lambda} - \lambda_0|/\lambda_0 \times 100\% =$

(2) Observe the structure of "double-line" of sodium yellow light, calculate the wavelength difference and compare it with 6 Å.

3. Discussion

Part 2 Topics of Experiments

2.24 Study Material Dispersion of a Prism

In the previous experiment "Measuring the wavelengths of light with a transmission diffraction grating", we have learned how to operate the grating spectrometer. In this experiment, we will study the phenomenon of dispersion in which the refractive index of the material of a prism depends on the wavelength of the incident ray. We will measure the refractive index of a prism at different wavelengths using the prism spectrometer, and calculate the coefficients in Cauchy's equation.

1. Purpose

(1) Describe the operation of a prism spectrometer;

(2) Measure the apex angle of the prism, and the refractive index of the prism material with the method of <u>minimum deviation angle</u>;

(3) Study the relationship between the refractive index of a glass prism and the wavelength of light;

(4) Calculate the coefficients of Cauchy's equation with least square method.

2. Principle

(1) Refraction and dispersion of the prism material.

When a beam of monochromatic light obliquely impacts on an interface between two transparent materials, it will be refracted and deviate from its original direction (Fig. 2.24.1), which follows the Snell's law:

$$n_1 \sin \theta_1 = n_2 \sin \theta_2 \qquad (2.24.1)$$

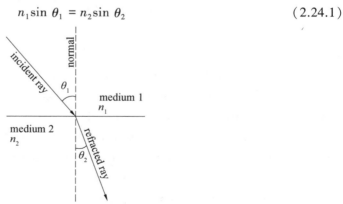

Fig. 2.24.1 Refraction of light at the interface between two media ($n_2 > n_1$)

where θ_1 and θ_2 are the angles of incidence and refraction; n_1 and n_2 are the refractive index of medium 1 and medium 2 respectively.

When a beam of polychromatic light (e.g. white light) falls obliquely on the surface of a dispersive prism and passes through it, the material dispersion (the property of a medium that the refractive index is wavelength-dependent) causes white light to be spread out into a spectrum of colors. This phenomenon is called dispersion (Fig. 2.24.2). It should be mentioned that the speed of light is the same for all wavelengths in vacuum, but in a dispersive medium, the speed of light is different for different wavelengths. That means different wavelengths of light will be refracted at different angles, and the refractive index will be different for different wavelengths in a dispersive medium. The relationship between the refractive index n and the wavelength of light λ for a normal dispersive material may be presented by so-called Cauchy's equation:

$$n = A + B\frac{1}{\lambda^2} + C\frac{1}{\lambda^4} + \cdots \quad (2.24.2)$$

where A, B and C are coefficients that can be determined for a specific material. Usually the first two terms may give an accuracy enough to express the dispersion. By using the experimental data of refractive indices for the given wavelengths, the coefficients in the equation can be fitted.

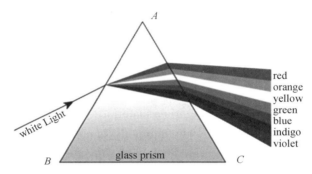

Fig. 2.24.2   Dispersion in a prism

(2) The angle of minimum deviation and the calculation of the refractive index of a prism material.

A monochromatic (single color or wavelength) parallel light beam from air is obliquely incident onto the surface AB of a transparent prism, refracted into the prism, and emerges from the surface AC. As shown in Fig. 2.24.3, the light is refracted and deviates from its original direction. The angle θ between the direction of the incident beam and the emergent one is called the angle of deviation, it is different for each color or wavelength, the red component has the longest wavelength, and it deviates the least. It can be proved that the angle of deviation is least when the angle of incidence i and the angle of emergence i' are equal. The angle of minimum deviation $\theta_{\min}$ and the prism apex angle α are related to the refractive index of the prism through Snell's law. The relationship is given by

$$n = \frac{\sin\dfrac{\alpha + \theta_{\min}}{2}}{\sin\dfrac{\alpha}{2}} \quad (2.24.3)$$

Then the index of refraction n can be calculated when apex angle α and the angle of minimum deviation are measured.

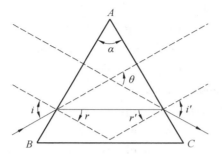

Fig. 2.24.3   The angle of minimum deviation

(3) Measurement of the apex angle α of a prism.

Put the prism on a proper position of the spectrometer table so that the incident light from the collimator could be successfully split by the apex, reflected by two faces and received by the telescope (Fig. 2.24.4). Adjust the prism so that the apex angle α splits the beam into nearly equal parts. Let φ_L and φ_R be the angular positions of two reflected beams, the difference in angular positions (the angle between two beams) is twice the apex angle α, given by

$$\alpha = \frac{1}{2} |\varphi_L - \varphi_R| \tag{2.24.4}$$

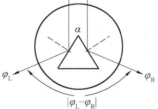

Fig. 2.24.4   Measurement of the prism apex angle α

Record the readings from scales on each side, find the corresponding apex angles ($\alpha_I = \frac{1}{2} |\varphi_{LI} - \varphi_{RI}|$ and $\alpha_{II} = \frac{1}{2} |\varphi_{LII} - \varphi_{RII}|$), here I and II represent the reading windows of both sides respectively, and average them to eliminate any systematic error from misalignment of the circle scale with respect to bearing axis. The apex angle α is given by

$$\alpha = \frac{1}{4} [|\varphi_{LI} - \varphi_{RI}| + |\varphi_{LII} - \varphi_{RII}|] \tag{2.24.5}$$

3. Procedure

(1) Adjust the spectrometer.

Put the mirror on the spectrometer stage and adjust the spectrometer to meet these requirements (details see Part A of experiment: measuring the wavelengths of light with a transmission diffraction grating):

①The telescope is focused at infinity and the optical axis of the telescope is perpendicular to the central axis of spectrometer strictly.

②The collimator emits (emerges) the parallel light and the optical axis of the collimator is perpendicular to the central axis of spectrometer strictly.

(2) Measure the prism apex angle α.

Take down the mirror and put the prism on the spectrometer table, as shown in Fig. 2.24.4. Find the reflected image of the slit from either side of the prism angle α by rotating the telescope, and record the angular positions for each reflected image. Repeat this procedure 5 times, and complete Table 2.24.1 in the experimental report. Calculate the angle α by Eq.(2.24.5).

(3) Measure the angle of minimum deviation.

①Rotate the prism to a position as shown in Fig. 2.24.3, and with the unaided eye, locate the emergent spectrum lines. Move the telescope in front of the eye and examine the spectrum lines.

②Rotate the prism back and forth slightly, observe the direction of motion of the yellow light (578.0 nm), and note the reversal of the direction when the prism is rotated in one direction. Stop the prism at the position of the reversal of motion of the yellow component of the spectrum and rotate the telescope to aim at the line. This is the position for minimum deviation of this component, record the corresponding scale readings $\theta_{\text{refraction}}$ (in Ⅰ and Ⅱ reading window).

③Observe other components (546.1 nm, 491.6 nm, 435.8 nm, 404.7 nm), repeat the step ②, measure the corresponding angle of minimum deviation one by one.

④Remove the prism, rotate the telescope towards the collimator, find the image of slit, record the angular position of incident angle $\theta_{\text{incidence}}$, complete Table 2.24.2 in the Experimental report.

(4) Based on the data in Table 2.24.2 in the report, calculate the coefficients of the first two term A and B in Eq.(2.24.2) $\left(n = A + B\dfrac{1}{\lambda^2}\right)$, by using the graphic paper or by the least square method.

4. Discussion

(1) Is the refractive index of a dispersive material the same for all wavelengths? Explain this in terms of the speed of light in the material.

(2) Which color of visible light has the greatest dispersion in this experiment?

Part 2　Topics of Experiments

Experimental Report

Study Material Dispersion of a Prism

Name:　　　　　　　　Student ID No.:　　　　　Score:

Date:　　　　　　　　　Partner:　　　　　　　　Teacher's signature:

1. Purpose

2. The experimental data and processing

(1) Measure the prism apex angle α (Table 2.24.1).

Table 2.24.1　Measure the prism apex angle α

No.		1	2	3	4	5
φ_L	I					
	II					
φ_R	I					
	II					
$\alpha = \frac{1}{4}[\,|\varphi_{L\,I} - \varphi_{R\,I}| + |\varphi_{L\,II} - \varphi_{R\,II}|\,]$						

(2) Measure the angle of minimum deviation (Table 2.24.2).

Table 2.24.2　Measure the angle of minimum deviation

λ/nm	$\theta_{\text{refraction}}$		$\theta_{\text{incidence}}$		$\theta_{\min}$		$\bar{\theta}_{\min}$	n
	I	II	I	II	I	II		
578.0 (Yellow)								
546.1 (Green)								
491.6 (Blue)								
435.8 (Indigo)								
404.7 (Violet)								

(3) Based on the data in Table 2.24.2, calculate the coefficients A and B for the first two terms in Eq.(2.24.2) $\left(n = A + B\dfrac{1}{\lambda^2}\right)$, with the least square method (Table 2.24.3).

Table 2.24.3 Calculation of the coefficients A and B using the least square method

λ/nm	$x = 1/\lambda^2$	$y = n$	x^2	xy
578.0 (Yellow)				
546.1 (Green)				
491.6 (Blue)				
435.8 (Indigo)				
404.7 (Violet)				
Average	$\bar{x} =$	$\bar{y} =$	$\overline{x^2} =$	$\overline{xy} =$

Notes:

For $y = kx + b$, we have

$$\begin{cases} \bar{x} = \dfrac{1}{n}\sum_{i=1}^{n} x_i \\ \bar{y} = \dfrac{1}{n}\sum_{i=1}^{n} y_i \\ \overline{x^2} = \dfrac{1}{n}\sum_{i=1}^{n} x_i^2 \\ \overline{xy} = \dfrac{1}{n}\sum_{i=1}^{n} x_i y_i \end{cases} \Rightarrow \begin{cases} k = \dfrac{\bar{x} \times \bar{y} - \overline{xy}}{(\bar{x})^2 - \overline{x^2}} \\ b = \bar{y} - k\bar{x} \end{cases}$$

So $\begin{cases} B = k = \dfrac{\overline{xy} - \bar{x}\bar{y}}{(\bar{x})^2 - \overline{x^2}} = \\ A = b = \bar{y} - k\bar{x} = \end{cases}$

(4) Based on the data in Table 2.24.2, calculate the coefficients A and B for the first two terms in Eq.(2.24.2) $\left(n = A + B\dfrac{1}{\lambda^2}\right)$. Plot the experimental data on a graph of n versus $\dfrac{1}{\lambda^2}$ and calculate the slope of the line with graphic method.

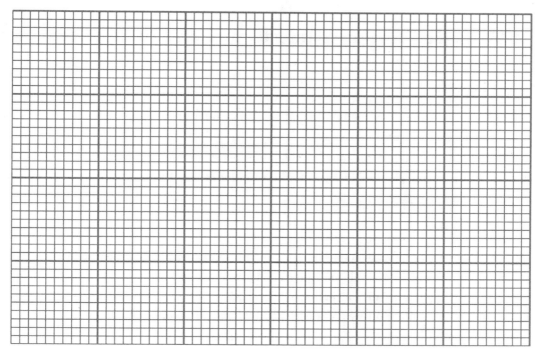

3. Discussion

2.25 Equal Thickness Interference and Its Applications

1. Purposes

(1) To observe the phenomena of equal thickness interference and understand the wave nature of light.

(2) To learn the interferometric method for measuring ① the curvature radius of a spherical lens by using Newton's rings; ② the tiny thickness of a thin body by using optical wedge interference.

2. Principle

Light interference reveals its wave nature. The interference phenomena may find wide applications in scientific research and interferometric engineering, such as the measurement of optical wavelength, surface inspection in flatness, sphericity, and smoothness, accurate measurement of a length and tiny deformation, and so on. The equal thickness interference is an important and very common phenomenon of interference.

Consider a homogeneous thin film with refractive index n and thickness d, on which a beam is incident with an angle α. The multiple reflection and refraction at the both surfaces form two set of rays: the reflected ones and transmitted ones (Fig. 2.25.1). The difference in optical length between the first reflected beam and the adjacent beam can be calculated by

$$\Delta = 2d\sqrt{n^2 - n_1^2 \sin^2\alpha} + \frac{\lambda}{2} \qquad (2.25.1)$$

where n_1 is the refractive index of the medium from which the beam is launched in. Here a half wave loss resulting from reflection of light from rare to dense medium is considered by adding a term of $\lambda/2$. Sometimes for convenience the optical length difference can be expressed by

$$\Delta = 2nd\cos\beta + \frac{\lambda}{2} \qquad (2.25.2)$$

where β denotes the inner refraction angle in the medium. This optical length difference decides the interference pattern of the reflected rays at P (the pupil). If the beam is normally incident, $\alpha = \beta = 0$, we have

$$\Delta = 2nd + \frac{\lambda}{2} \qquad (2.25.3)$$

If the optical length difference is the odd times of half wavelength, namely

$$\Delta = 2nd + \frac{\lambda}{2} = (2m + 1)\frac{\lambda}{2}, \quad m = 0, 1, 2, 3,\ldots \qquad (2.25.4)$$

all the reflected rays will interfere destructively and result in minimum intensity forming a dark fringe. If the optical length difference is the even times of half wavelength, then

$$\Delta = 2nd + \frac{\lambda}{2} = m\lambda, \quad m = 1, 2, 3,\ldots \qquad (2.25.5)$$

Part 2   Topics of Experiments

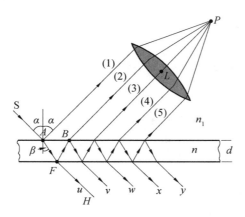

Fig. 2.25.1   The reflection and refraction on
the surfaces of a thin film

it will lead to a constructive interference, and a bright fringe can be seen in the reflection.

It can be seen that the fringe order m corresponds to the thickness of the film d, and the fringes describe the contour lines of the film, so this interference has a name of "equal thickness interference".

(1) The Newton's rings.

The equal thickness interference can be formed by using a convex lens and a glass plate (see A and B in Fig. 2.25.2) and measured by using the reading microscope. The observation are usually made at normal incidence as shown in Fig. 2.25.2, where the glass plate G is used as a beam splitter to reflect the light to A and B. The reflected light from B passes through A and G and imaged in the microscope T.

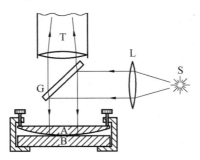

Fig. 2.25.2   The setup for observation and measurement of Newton's rings

Since the fringes are produced in the air film between the convex surface of the plano-convex lens and the surface of plane glass, the contour lines are circular by symmetry. These ring-shaped fringes were studied in detail by Newton, but he didn't explain it correctly.

If we assume that A and B are just touching at the center, the curvature radius of the surface A is R, the thickness of the air film is d, and the radius of the mth ring is r_m (Fig. 2.25.3(a)), then we have

$$R^2 = (R - d)^2 + r_m^2$$

Since $d \ll R$, we have
$$r_m^2 = 2Rd$$
namely
$$d = \frac{r_m^2}{2R} \qquad (2.25.6)$$

Applying the condition for dark fringes and assuming the refractive index of air $n = 1$, the radius of the mth dark fringe can be calculated:
$$r_m = \sqrt{mR\lambda} \qquad (2.25.7)$$

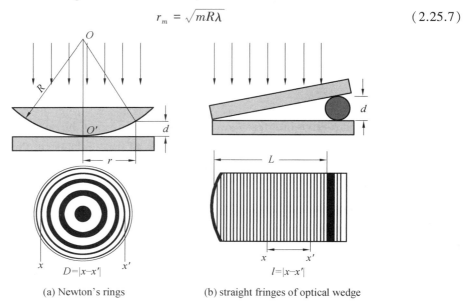

(a) Newton's rings   (b) straight fringes of optical wedge

Fig. 2.25.3   The comparison of Newton's rings and straight fringes of optical wedge

In the real case the contact of plano-convex lens and flat glass at the center are not at geometric point, which results in an uncertainty in determining the order of fringes near the center. The fringe at very center sometimes appears in form of bright spot or dark spot. It is difficult to measure the radius r_m and determine the absolute order of a fringe m. So in practice the measurement of the radius r_m is replace by that of diameter D_m, then
$$D_m = 2\sqrt{mR\lambda} \qquad (2.25.8)$$

Furthermore, in order to minimize the errors the measurement is made repeatedly and the data is processed by the method of successive difference. It is not necessary to judge the absolute order anymore, only need to measure the relative order m, k and corresponding diameter D_m, D_k. Thus the curvature radius of the convex surface can be expressed by
$$R = \frac{D_m^2 - D_k^2}{4(m-k)\lambda} \qquad (2.25.9)$$

(2) Interference of the optical wedge formed by two flat surfaces.

When a thin object (say a piece of hair or tape) is inserted along the edges of two pieces of flat glass, it forms an air film (optical wedge) between the two surfaces. If a monochromatic beam is normally irradiated on the surfaces the equal thickness interference occurs, and the fringes are usually straight lines (Fig. 2.25.3(b)). Assuming that the refractive index of air is $n = 1$, the interference condition for dark fringes becomes

$$\Delta = 2d + \frac{\lambda}{2} = (2m + 1)\frac{\lambda}{2}, \quad m = 0, 1, 2, 3, \ldots \quad (2.25.10)$$

where the thickness of air film d corresponding to the order m is

$$d = m\frac{\lambda}{2} \quad (2.25.11)$$

From above expression, $m = 0$ corresponds to $d = 0$, that means that the fringe is dark at the contact point.

If we read out the total number of fringe order N by counting fringes from the contact point to the edge where the thin object is laid, the thickness of the body can be calculated by

$$d = N\frac{\lambda}{2} \quad (2.25.12)$$

since $N = m$. In practice, the total number N may be obtained by measuring the number of fringes (h) in unit distance and the total distance L between the contact point and the wedge of the thin object. h can be calculated from the distance for 10 fringe interval l by $h = 10/l$. So

$$d = Lh\frac{\lambda}{2} \quad (2.25.13)$$

3. The contents of experiment

(1) Measure the curvature radius of a plano-convex lens R.

①Be familiar with the method to form Newton's rings, observe the characteristics of interference fringes from a microscope: the shape, the diameter variation in adjacent rings, the status of central fringe (bright or dark?), and analyze the variations in fringes when slightly press the lens from above.

②Measure the diameters corresponding to 11th – 20th ring by using of the measuring microscope with micrometer. In order to eliminate the errors from screw pitch, it is not allowed to change the rotation direction during the measurement.

(2) Measure the thickness of a tape.

①Form an optical wedge using two glass plates and a piece of recording tape, adjust the direction of the tape and slightly press the upper plate to let the fringe direction parallel to the tape direction. All the directions should be vertical to the measuring direction.

②Repeatedly (5 times) measure the total distance between the contact point and the tape L.

③Repeatedly (5 times) measure the intervals corresponding to 10 fringes l at different places.

4. Discussion

(1) Compare the similarity and difference between the Newton's rings and the straight fringes of the optical wedge.

(2) Please draw the fringe when there is a bulge or dent on the upper surface of the second glass plate in the optical wedge experiments.

Experimental Report

Equal Thickness Interference and Its Applications

Name:　　　　　　　　Student ID No.:　　　　　　Score:

Date:　　　　　　　　　Partner:　　　　　　　　　Teacher's signature:

1. Purpose

2. The experimental data and processing

(1) Experiment for Newton's rings (Table 2.25.1).

Table 2.25.1　Measure the Newton's rings at λ = 589.3 nm

Number of order		20	19	18	17	16
Positions of ring	left					
	right					
Diameter D_m/mm						
Number of order		15	14	13	12	11
Positions of ring	left					
	right					
Diameter D_k/mm						
$D_m^2 - D_k^2$/mm^2						

The mean value of curvature radius R can be calculated by

$$\bar{R} = \frac{\overline{D_m^2 - D_k^2}}{4(m-k)\lambda} =$$

The uncertainty of $D_m^2 - D_k^2$ can be calculated by

$$U_{D_m^2-D_k^2} = \sqrt{S_{D_m^2-D_k^2}^2 + u^2} \approx S_{D_m^2-D_k^2} = \sqrt{\frac{\sum_{i=1}^{5}[(D_m^2-D_k^2)_i - \overline{D_m^2-D_k^2}]^2}{5(5-1)}} =$$

So the resultant uncertainty for R is

$$U_R = \frac{U_{D_m^2 - D_k^2}}{4(m-k)\lambda} =$$

The results can be expressed as

$$\begin{cases} R = \bar{R} \pm U_R = \\ E = \dfrac{U_R}{\bar{R}} \times 100\% = \\ \text{Confidence probability } P = 68.3\% \end{cases}$$

(2) Experiment for optical wedge (Table 2.25.2).

Table 2.25.2 Measure the thickness of the tape

L_1	L_2	L/mm	$\bar{L}/\text{mm}$	l_1	l_2	l/mm	$\bar{l}/\text{mm}$	h/mm^{-1}

$d = Lh\lambda/2 =$

3. Discussion

2.26 Polarized Light: Generation and Determination

1. Purpose

(1) To understand the definition of natural light and polarized light;
(2) To verify the Malus's law;
(3) To learn how to acquire and identify polarized light using polarizer and wave plate.

2. Principle

Polarization is one of the most important properties to reveal the wave nature of light. Many interesting optical phenomena that are related with polarization have been extensively used in practice.

(1) Natural light and polarized light.

As part of the electromagnetic wave, light is a <u>transverse wave</u>, whose vibration directions of the electric and magnetic field (E and B) vectors are perpendicular to the propagation direction (Fig. 2.26.1). If the vibration vector is confined to a certain direction, the light is called <u>linearly polarized light</u> or <u>plane polarized light</u>. The vibration direction of the electric field (E vector) and the propagation direction define the polarization direction and polarization plane. As shown in Fig. 2.26.1, the polarization direction is the Y axis direction and the polarization plane is the YC plane.

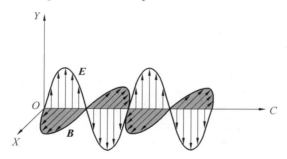

Fig. 2.26.1　The illustration of an linearly polarized electromagnetic wave

Generally speaking, the light radiated from a molecule or an atom is linearly polarized light (Fig. 2.26.2(a)). However, light from an ordinary source consists of waves emitted from a large number of atoms or molecules. Each atom/molecule produces a wave whose polarization direction is determined by the orientation of the atomic/molecular vibration. With lots of atoms/molecules in the source the emitted light is <u>unpolarized</u> because the vibration vectors are randomly oriented and all the polarization directions have equal probability. This unpolarized light is called <u>natural light</u> (Fig. 2.26.2(b)). If for some reason the E vectors become preferentially oriented, the light is <u>partially polarized</u> as shown in Fig. 2.26.2(c).

Besides the linearly polarized light, there are other kinds of polarized lights whose vibration direction may change with time. If the **E** vector end makes an elliptic or a circular trace in the cross section vertical to the propagation direction the light is called <u>elliptically polarized light or circularly polarized light</u>.

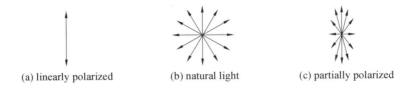

(a) linearly polarized   (b) natural light   (c) partially polarized

Fig. 2.26.2   The polarization as denoted by the **E** vector viewed along the propagation direction

(2) How to acquire polarized light.

The polarized light can be acquired by several ways: refraction, reflection, selective absorption, and scattering.

①Polarization by Refraction. Polarization by refraction generally refers to the <u>birefringence</u> of some crystals. In an optically isotropic medium, such as glass, light travels with the same speed in all directions. As a result, the material is characterized by a single index of refraction. However, in some anisotropic crystals the speed of light is not the same in all directions. For example, the calcite crystal exhibits birefringence, which is characterized by two indices of refraction. Unpolarized light passing through this crystal is separated into two components or beams, both of them are linearly polarized as shown in Fig. 2.26.3. One beam is the <u>ordinary (o) light</u>, and the other is the <u>extraordinary (e) light</u>. When we observe something through a calcite crystal, we can see the double images (Fig. 2.26.3).

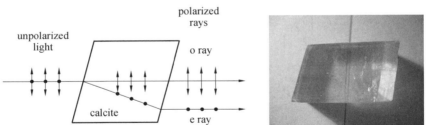

Fig. 2.26.3   The principle and observation of birefringence

The polarizer and analyzer used in this experiment are Glan-Thomson prisms which are fabricated with a piece of glass and a piece of Calcite crystal based on the birefringent principle. As shown in Fig. 2.26.4, the interface of the glass and Calcite splits the incident natural light into o beam (vertical polarized component) and e beam (parallel polarized component). The o beam can transmit the Calcite directly since the refractive index of this beam (n_o = 1.658 4) is a little larger than that of glass (n_g = 1.655). The e beam deflected to another direction at the interface by the total internal reflection since the refractive index for this beam (n_e = 1.486 4) is less than that of glass (n_g = 1.655).

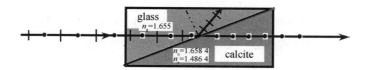

Fig. 2.26.4　Glan-Thomson prism

②Polarization by Reflection. When light is incident on a material such as glass, part of the light is reflected and another part is transmitted. The reflected light is usually partially polarized, and the degree of polarization depends on the angle of incidence. For 0° and 90° incidence angles, the reflected light is unpolarized. However, for intermediate angles, the light is polarized to some extent. Complete polarization occurs at an optimum angle called the polarization angle θ_p or critical angle θ_c (Fig. 2.26.5). This happens when the reflected and refracted beams are 90° apart, and θ_p (or θ_c) is specific for a given material.

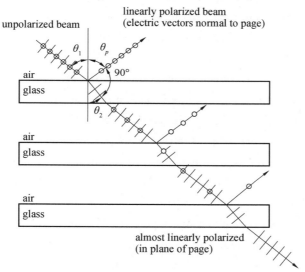

Fig. 2.26.5　Polarization by reflection and transmission

Referring to Fig. 2.26.5, since $\theta_1 = \theta_p$, then $\theta_1 + \theta_2 = 90°$ or $\sin \theta_2 = \sin (90° - \theta_1) = \cos \theta_1$. By Snell's law, there is

$$\frac{\sin \theta_1}{\sin \theta_2} = \frac{\sin \theta_1}{\cos \theta_1} = \tan \theta_1 = n \tag{2.26.1}$$

where n is the index of refraction of the glass. Thus

$$\tan \theta_p = n \tag{2.26.2}$$

This expression is called Brewster's law, and θ_p is called Brewster's angle.

It can be seen from Fig. 2.26.5 that the reflected beam is horizontally polarized at Brewster's angle. Sunlight reflected from water and metallic surfaces (for example, from a car) is partially polarized. If the surface is horizontal, the reflected light has a strong horizontal component. This is used in polarizing sunglasses. The transmission axis of the glasses is vertically oriented so as to absorb the reflected horizontal component and reduce the light intensity. Also notice that the refracted beam is partially polarized. If a stack of glass plates are used, the transmitted beam becomes more linearly polarized as shown in Fig. 2.26.5.

③Polarization by selective absorption. Some birefringent crystals absorb one of the polarized components more than the other. This selectively absorbing property is called dichroism. A typical dichroic crystal is quinine sulfide periodide (usually called herapathite) whose special polarizing properties were discovered by a British physician (W. Herapath) in 1852. This crystal has been extensively used for making polarizers. Around 1930, Edwin H. Land, an American scientist, found a way to align tiny dichroic crystals in sheets of transparent celluloid. A thin sheet of polarizing material named polaroid can be produced in this way. Then, polymer materials were used to produce better polarizing films. This kind of film is stretched in the manufacturing process so as to align the long molecular chains of the polymer. With proper treatment, the valence electrons of the molecules can move along the chains, and therefore the molecules absorb light with the E vectors parallel to the chains and transmit light with the E vectors perpendicular to the chains. The direction perpendicular to the chains is usually called the transmission axis or polarization direction.

④Polarization by scattering. Scattering is such a process that a medium absorbs light and then reradiates. For example, if light is incident on gas, the electrons in the gas atoms or molecules can absorb and reradiate (scatter) part of the light. The electrons absorb the electric field component of the light wave. An electron works as a small antenna, it radiates light in all directions except its vibration axis. Hence the scattered light is partially polarized.

(3) The Malus' law.

When unpolarized light passes through a polarizer (usually made from crystal or polymer sheet), polarized light is obtained. The polarization of light may be analyzed (identified) by another polarizer, which acts as an analyzer. The magnitude of the E vector parallel to the transmission axis of the analyzer is $E_0 \cos \theta$ as shown in Fig. 2.26.6. Since the intensity is the square of the amplitude, the intensity of the transmitted light through the analyzer is

$$I = I_0 \cos^2 \theta \qquad (2.26.3)$$

where I_0 is the maximum intensity of light through the analyzer and θ is the angle between the transmission axis of the polarizer and analyzer. This expression is named Malus' law after a French physicist E. I. Malus who discovered it. If $\theta = 90°$, we have a condition of cross-polarizers, and no light is transmitted through the analyzer.

(4) Wave plate and its functions.

One can employ the birefringence to make wave plates. When a linearly polarized beam incident along the direction perpendicular to an axis of some anisotropic crystal the o beam and e beam may propagate through it in the same direction but with different phase velocity, which leads to a phase difference:

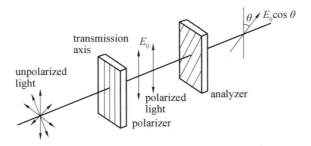

Fig. 2.26.6   Polarizer and analyzer

$$\Delta\varphi = 2\pi(n_o - n_e)\frac{d}{\lambda} \qquad (2.26.4)$$

where d denotes the thickness of crystal, $n_o - n_e$ represents the difference in the refractive indices of the o and e beams at the wavelength λ. Designing proper thickness d one can make a phase difference $\Delta\varphi = 2m\pi + \pi/2$ corresponding to the optical path difference $m\lambda + \lambda/4$ (m is the integer), in such a case a $\lambda/4$ (quarter λ) wave plate is obtained. In the same way one may also get a $\lambda/2$ (half λ) wave plate that produces $\Delta\varphi = 2m\pi + \pi$.

The $\lambda/4$ wave plate may be used to generate an elliptically or circularly polarized beam from a linearly polarized one. As shown in Fig. 2.26.7(a), when the angle between the crystal axis and the polarization direction of the incident beam is $\alpha = \pi/4$ the decomposed o and e components are equal in amplitude and generate circularly polarized beam. The linearly polarized light remains linearly polarized when passed through a $\lambda/2$ wave plate. But the polarization direction is rotated by 2α as shown in Fig. 2.26.7(b).

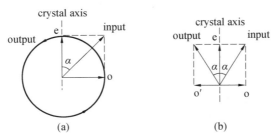

Fig. 2.26.7   The effects of $\lambda/4$ and $\lambda/2$ wave plates on the linearly polarized light

3. Experiment contents

In the experiment, the process controlling and data collection is performed by software on the PC. The experiment setup is shown on the posters. Before the experiment, the teacher will explain the function of the instruments. It is suggested that students put the laser and the detector correctly at first and then align the other optics.

(1) Verify Malus's law.

①Insert a polarizer between the laser and detector first, rotate the polarizer to make sure you get a higher polarized output and avoided saturation in the detector. Never rotate this polarizer in the following steps.

②Insert an analyzer between the polarizer and detector, rotate the analyzer using the software in the PC and observe the curve acquired by the detector. Then use the "angle searching" function to assign the zero position of the polarizer and to perform the following measurements.

③Rotate the analyzer from the zero position and read the data on the curve acquired by the detector. Fill in Table 2.26.1 in the experimental report and draw the $I/I_0 - \theta$ curve.

④Draw the $I/I_0 - \cos^2\theta$ curve using the 0° – 90° data to verify Malus's law.

(2) Study the functions of $\lambda/4$ wave plates.

①In the previous setup, adjust the orientation of the analyzer to the extinction position. Insert the $\lambda/4$ wave plate between the polarizer and analyzer. Rotate the wave plate and keep the analyzer in the extinction position. In some orientation, it seems the wave plate doesn't work, which indicates that the crystal axis of the wave plate is parallel with the polarization plane of the polarizer.

②Using the function of "angle searching" to fix the initial angle of the wave plate to 0°, 30° and 45°, and rotate the analyzer to measure the transmission vs rotation angle. Record the data in Table 2.26.2 in the report.

③Draw the three curves in one figure and analyze what happens when the linearly polarized light pass through the $\lambda/4$ wave plate and what information you can get from these curves.

Experimental Report

Polarized Light: Generation and Determination

Name:　　　　　　　Student ID No.:　　　　　　　Score:

Date:　　　　　　　Partner:　　　　　　　Teacher's signature:

1. Purpose

2. The experimental data and processing

(1) Verification of Malus's law (Table 2.26.1).

Table 2.26.1　The transmitted intensity after the analyzer

$\theta/(°)$	0	10	20	30	40	50	60	70	80	90
Intensity										
$\theta/(°)$	100	110	120	130	140	150	160	170	180	
Intensity										

The $I - \theta$ curve (0° – 180°).

The $I/I_0 - \cos^2\theta$ curve (0° – 90°).

Verify Malus's law (least squares or graphing methods).

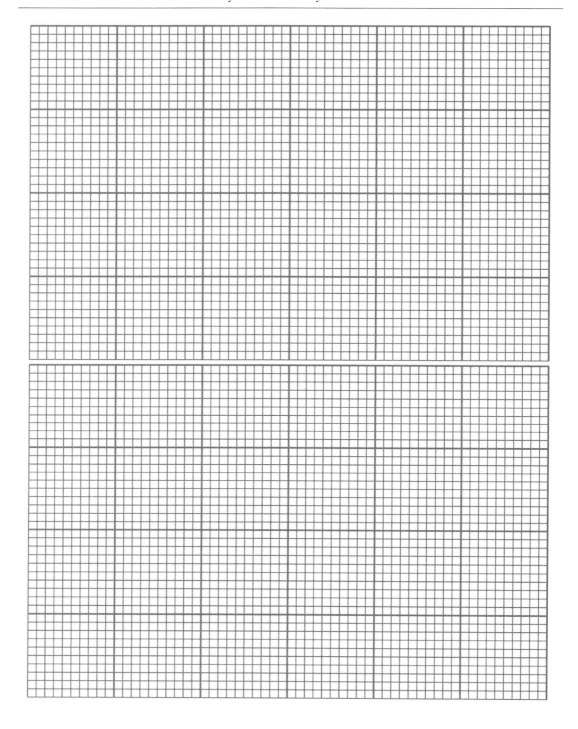

(2) The functions of λ/4 wave plate (Table 2.26.2).

Table 2.26.2 The transmitted intensity after the λ/4 wave plate and the analyzer

θ/(°)	0	10	20	30	40	50	60	70	80	90
0°										
30°										
45°										
θ/(°)	100	110	120	130	140	150	160	170	180	...
0°										
30°										
45°										

The $I - \theta$ curves ($\alpha = 0°$, $30°$, and $45°$).

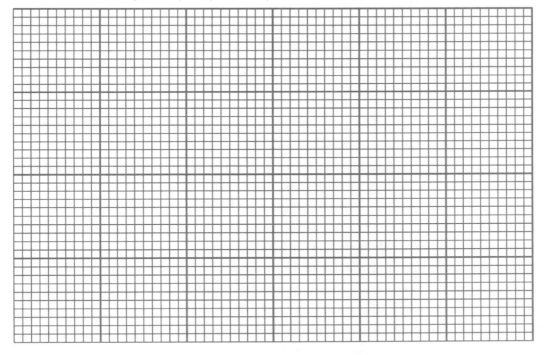

3. Discussion

(1) What kinds of polarized light can you get when the linearly polarized light pass through a λ/4 wave plate?

(2) How to distinguish the circularly polarized light and the natural light?

2.27 Holography Technique

The idea of holography was firstly proposed by the Hungarian-British physicist Dennis Gabor in 1948. His work was built on pioneering work in the field of X-ray microscopy by other scientists. This technique as originally invented was used in electron microscopy, which is known as electron holography. The optical holography was realized until the development of the laser in 1960. Dennis Gabor was awarded the Nobel Prize in Physics in 1971 "for his invention and development of the holographic method". Nowadays, holography is widely used in lots of fields not only in science laboratories but also in our daily life, such as dynamic holography, interferometric microscopy, data storage, art, security and so on.

1. Purpose

(1) To master the principle of optical holography;
(2) To master the method of designing and building the optical setup;
(3) To learn the method of recording a hologram and reconstructing and viewing the holographic image.

2. Principle

(1) A brief description of holography.

Holography is a technique that enables a light field to be recorded and later reconstructed when the original light field is no longer present, due to the absence of the original objects. Generally speaking, when we see a three-dimensional (3D) object around us, that means a light source is scattered off the object and the scattered light comes into our eyes. The information in the scattered light field includes amplitude, frequency and phase, which determine the brightness, color and distance of the object. The ordinary image formed by a lens (or camera) can record the amplitude and frequency of the scattered light field, and therefore we can see only the two-dimensional (2D) image. However, a hologram is a photographic recording all the information in the light field, and it is used to display a fully 3D image of the holographed subject, which is seen without the aid of special glasses or other intermediate optics.

(2) Recording a hologram.

The principle of recording a hologram is interference. Fig. 2.27.1 shows a typical setup for recording a hologram. A He-Ne laser is used as the light source that produces coherent light with at 632.8 nm. The shutter at the exit of the laser is used to control the exposure time. The beam splitter separates the laser into two beams, the first beam illuminates the object (the object beam) and another beam illuminates the recording medium directly (the reference beam). Two lenses in these two beams are used as beam expander to expand the illuminating areas. If the optical path difference between object and reference beams is shorter than the coherence length of the laser, the reference light and the light scattered from the object onto the recording medium can interference and form an interference pattern on the recording medium (photographic plat).

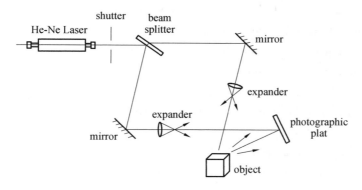

Fig. 2.27.1 The setup for recording a hologram

It can be seen from the nature of optical interference that the interference pattern of two single wavelength beams is determined by the amplitude and phase of these two beams. The electric field wave functions of the object and reference light are expressed as $E_o(t) = E_o\cos(\omega t + \varphi_o)$ and $E_r(t) = E_r\cos(\omega t + \varphi_r)$, where ω is angular frequency, $E_o(E_r)$ and $\varphi_o(\varphi_r)$ are amplitude and phase of the object (reference) light, respectively. The intensity of the interference light can be expressed as

$$I = E_o^2 + E_r^2 + 2E_oE_r\cos(\varphi_o - \varphi_r) \quad (2.27.1)$$

Therefore, the interference pattern on the photographic plat records both the amplitude and phase information of the scattered light from the object.

If the object and reference lights illuminating on a small region of the photographic plat are considered as collimated beams, the spacing of the interference fringe can be expressed as

$$d = \frac{\lambda}{2\sin\frac{\theta}{2}} \quad (2.27.2)$$

where λ is the wavelength of the laser and θ is the angle between the reference and object beams. The scattered light from a given point of the object illuminates different positions of the photographic plat with different incidence angles, thus the orientation and spacing of the interference fringes are different at different positions. On the other hand, the scattered light from any point of the object can illuminate a given position on the photographic plat with different incidence angles, thus the orientation and spacing of the interference fringes at a given position are also different. For these reasons, it seems that the interference fringes on the photographic plat are confusing, but in fact, they record the amplitude and phase of the reference and object lights.

Since the wavelength of light is of the order of 0.6 μm, it can be seen that very small changes in the optical paths travelled by either of the beams in the holographic recording system lead to movement of the interference pattern and reduce the visibility of the interference fringe. Such changes can be caused by relative movements of any of the optical components or the object itself, and even by local changes in air-temperature. If we want torecord a clear hologram, the light source, the optical components, the recording medium, and the object must all remain perfectly motionless relative to each other during the exposure, to within about a quarter of the wavelength of the light. For these reasons, we need an environment which provides sufficient mechanical and thermal stability to make sure the interference pattern is stable during the recording time. The exposure time required to record the hologram should not be too long, since longer exposure time needs more stable environment.

Another very important parameter is the intensities of the two beams. It can be seen from Fig. 2.27.1 that the reference beam is the expanded laser beam and it illuminates the photographic plat directly, and the object beam is the scattered light of the laser beam from the object. In general, the object has optically rough surfaces used in holography so that they scatter light over a wide range of angles. That means the intensity of the object beam is usually weaker than that of the reference beam. However, it is well known that the visibility of the interference fringes is determined by the relative intensity of the two coherence beams, the maxim visibility can be realized when the two beams have the same intensity. Therefore, we usually need to reduce the intensity of the reference beam and increase the intensity of the object beam in order to record more visible interference fringes and obtain clear hologram. When building the setup, we need to think about how to do that.

After exposure, the photographic plate needs developing, fixing, washing and drying procedures so as to exhibit and fix the hologram on it. Now, the photographic plate is ready for reconstructing and viewing.

(3) Reconstructing and viewing the holographic image.

The principle of reconstructing and viewing the holographic image is diffraction. As mentioned before, the hologram on photographic plat is interference fringes and we can't see the holographic image by directly observe under natural light. When a laser beam illuminates the photographic plat, the hologram's surface pattern acts as a grating and diffracts the incident light to produce a light field identical to the one originally produced by the scene and scattered onto the hologram. Since the interference fringes on the photographic plate records both the amplitude and phase information of the scattered light from the object, the diffracted light can also contain these information.

The sketch map for reconstructing and viewing the holographic image is shown in Fig. 2.27.2. The reconstructing beam illustrates the photographic plat in the same direction as the reference in recording procedure. When observing in the + 1 order direction, we can see the 3D virtual image that change realistically with any change in the relative position of the observer. The real image exhibits on a receiving screen in the − 1 order direction, and the 3D effects can be observed when moving the photographic plat and illuminating different positions of the interference pattern.

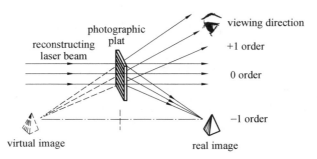

Fig. 2.27.2　Sketch map for reconstructing and viewing the holographic image

It is worth noting that the reconstructing beam illustrates the photographic plat in the same direction as the reference beam in recording procedure. Thus when recording the hologram, the angle between the reference and object beams should not be too small; otherwise the angle between the 0 and 1 (or −1) is too small, which makes it difficult to observe the holographic image. Of course, the angle between the reference and object beams should not be large. This is because the larger angle will produce smaller spacing of the interference fringe (Eq.(2.27.2)), which requires more rigorous stability of the environment and increases the difficulty of experiments.

(4) Characteristics of holography.

Holography is definitely different from ordinary photographic technique in recording, reconstructing and viewing. The characteristics of holography are summarized as follows.

①The holographic image is 3D since the hologram on the photographic plat record all the information in the light field of the scattered object beam. The Image changes realistically with any change in the relative position of the observer, and some positions on the object that can't be seen in one observing direction may be seen in other directions.

②The interference fringes at any area of the photographic plat are produced by the interference of the reference and object beams, and records the light information from all the positions of the object surface. So, if a holographic photographic plat is broken into pieces, any piece can reconstruct the whole image of the object.

③Because the principle of recording a hologram is interference, the interference fringes can be independently recorded in a single photographic plat for many times. After once exposure, if we alter the direction of the photographic plat or the reference beam, and change another object to record the hologram again on the same photographic plat, finally, the holographic images of different objects can be observed in different directions. This is the idea of holographic data storage. On the other hand, if we keep the photographic plat and the reference beam unchanged, and make a tiny (several micron) deformation on the object and record the hologram again, finally, the holographic image will show visible fringes at the deformation position. This is the idea of holographic interferometric microscopy.

④If the wavelength of the reconstructing laser is different from that of the recording laser, the holographic image can still be observed, but the image is different from the real object in both size and color. The image will be amplified (reduced) when a longer (shorter) wavelength laser is used for reconstructing.

⑤The brightness of the reconstructing image depends on the intensity of the reconstructing beam. The reconstructing image may be about 10^3 times bright than the real object if the reconstructing laser is intense enough.

3. Experiment contents

(1) Build the setup according to Fig. 2.27.1. The set up should meet the following conditions: the optical path difference between object and reference beams is shorter than 3 cm, the distance between the object and the photographic plate is about 10 cm, the angle between the reference and object beams is 30°-40°, the intensity ratio of the reference and object beams is 2∶1-10∶1, all the items are fixed. Use an exposed photographic plate to build the setup, a new photographic plate will be delivered before exposure.

<u>Notes: Keep your eyes from the laser beam and do not touch the surface of any optics.</u>

(2) Exposure. Turn off all the light sources except the He-Ne laser and the dark green lamp. Fix the new photographic plate in the mount. The exposure time is 10-15 s depends on the laser intensity.

<u>Notes: DO NOT WALK AND SPEAK during the exposure process.</u>

(3) Developing, fixing, washing and drying the photographic plate. The developing time is about 1 min until the plate becomes dark. After developing, wash the plate and then fixing. The fixing time is 4-5 min. After fixing, wash and dry the plate.

<u>Notes: Keeping the medium face upwards in the developing and fixing procedures to avoid damaging the medium.</u>

(4) View the holographic image.

After drying the photographic plate, we can view the holographic image as Fig. 2.27.2.

①View the virtual image, compare its size and position with the real object. Insert a beam expander in the laser beam to expand the illuminating areas on the photographic plate. Change the relative position of the observer to see the 3D effects. Cover part of the photographic plate and view the virtual image in the uncover position, the observer can still see the whole rather than part of the image.

<u>Notes: The clear virtual image can only be observed at specific direction (the + 1 order diffraction direction).</u>

②View the real image exhibit on a receiving screen.

Move away the beam expander and let the laser beam illuminate the photographic plate directly. Put a piece of white paper as receiving screen in the − 1 order direction as shown in Fig. 2.27.2. Change the angle of the photographic plate, the real image can be seen on the receiving screen. Move the photographic plate, let the laser beam illuminate different positions, the observer can see the 3D effects.

(5) Fill in Table 2.27.1 in the experimental report when finish all the experiment.

4. Discussion

(1) Please summarize the differences between holography and ordinary photographic techniques.

(2) What items are required to make a hologram? What should be paid attention to in the experiment?

Experimental Report

Holography Technique

Name:　　　　　　Student ID No.:　　　　　　Score:

Date:　　　　　　Partner:　　　　　　Teacher's signature:

1. Purpose

2. Procedure and notes of this experiment

3. Data recording

Draw the sketch map of your setup and record the experimental parameters in Table 2.27.1.

Table 2.27.1　The summarized parameters in the experiment

Optical path of the object beam	Optical path of the reference beam	Angle between object and reference beams	Developing time	Fixing time

4. Discussion

2.28 Michelson Interferometer

The Michelson interferometer plays an important role in the modern physics and modern metrology. A. A. Michelson and E. W. Morley used it to complete the famous Michelson-Morley "aether drift" experiment. The experiment became one of the experimental bases for denying the existence of aether and establishing Einstein's special theory of relativity. Michelson won the 1907 Nobel Prize in Physics for his invention of interferometers and measurements of the speed of light.

Michelson interferometer is designed with exquisite and versatile, its basic structure and design ideas, enlightening scientists, and laying the foundation for future generations to develop a variety of other forms of interferometer. The scientists also employ it to study the fine structure of the spectrum, using the wavelength of light to calibrate the length standard. The 2017 Nobel Prize-winning gravitational wave detection is also based on the basic principles of the Michelson interferometer.

1. Purpose

(1) To understand the structure and adjustment method of the Michelson interferometer;

(2) To observe the characteristics of the fringes of non-localized interference, localized interference (such equal-inclination interference, equal-thickness interference) and examine the conditions for their formation;

(3) To observe the characteristics of white light interference.

2. Principle

(1) Principle and structure of the instrument.

Fig. 2.28.1 is a schematic of the principle of a Michelson interferometer. M_2' is the virtual image of the plane mirror M_2 formed by the beam splitter BS semi-reflective glass plate. Observers from point of view O, not only the beam (1) is from the direction of M_1, the beam (2) also seems to be from the plane M_2'. Thus the interference fringes produced by the interferometer are similar to those produced by the air film formed by the plane M_2' and the plane M_1. The formation of the various interference fringes discussed below is considered in this way.

The compensation plate CP is a piece of material, the thickness is the same as the beam splitter BS, and the flat glass placed parallel to it, its role is to make the beam (2) and the beam (1) through the same thickness of the glass, so that the two beams in the glass experience the same optical path.

Part 2 Topics of Experiments

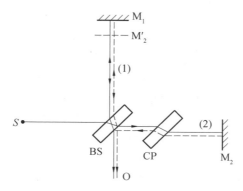

Fig. 2.28.1 Schematics of Michelson interferometer

Fig. 2.28.2 is the structure of the Michelson interferometer. The plane mirrors M_1 and M_2 are adjusted by the screws AS_1 and AS_2 behind them, respectively. Rotate the micrometer knobs so that they move parallel on the rails. The position of M_1 corresponds directly to the position of the micrometer, while M_2 is driven by a lever mechanism with a ratio of 1/20, so its real moving distance is twentieth the reading of the micrometer. Because the measurement accuracy of M_2 position is much higher than that of M_1, M_2 is used as a movable mirror during the experiment and M_1 as the fixed mirror.

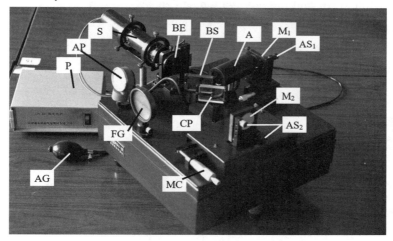

Fig. 2.28.2 The structure of a Michelson interferometer

M_1—Fixed mirror; M_2—Movable mirror; AS_1, AS_2—Adjustment screw; BS—Beam Splitter; CP—Compensation Plate; S—He-Ne laser; P—He-Ne laser power supply; BE—Beam expander; MC—spiral micrometer; A—gas chamber; AG—rubber ball; AP—barometer; FG—Frosted glass plate

Because of the two-arm separation structure, it is easy to insert a transparent component (such as a gas chamber or liquid tank for measuring their refractive index) in its optical path, and to measure certain physical parameters through changing the optical path difference. A mirror can also be mounted on a component to be measured, such as piezoelectric ceramics, to measure small changes in length.

(2) Interference fringes in the Michelson interferometer.

Both non-localized interference and localized interference may take place in a Michelson interferometer depending on the nature of the light source.

① Non-localized interference. When a monochromatic point light source irradiates the interferometer, at each point in the overlap of the two spherical waves reflected by the air film between plane M_1 and M'_2, there is an intersection of coherent light, and the overlap area extends from its surface to infinity. Interference fringes are everywhere in this vast area, so it is called "non-localized interference". The He-Ne laser used in the experiment can be seen as a monochromatic light. However, the laser beam is dispersed by a short focal length lens, or transmitted by optical fiber, which can be considered a point light source when it is coupled from the end of the fiber to the air. As shown in Fig. 2.28.3, any light of the point light source is reflected by the plane mirror M_1 and M'_2 respectively, which is equivalent to two virtual light sources emitting two beams of coherent light. The circular or oval interference fringes can be observed at different positions, and the shape of the stripes is related to the relative angle between M_1 and M'_2, the angle of exposure to the light source, and the angle of observation. Sometimes the center of the fringes cannot be seen. If the screen is perpendicular to $S'_1 S'_2$, the interference fringes will be a set of concentric circles, the center of which is the intersection of the extension line $S'_1 S'_2$ and the observation screen.

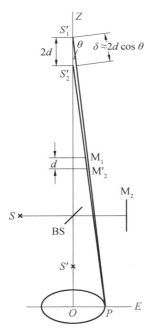

Fig. 2.28.3 Non-localized interference

When the distance between S'_2 and the screen $Z \ll 2d$, the optical path difference from S'_1 and S'_2 to a point P on the screen δ is

$$\delta \approx n_0 2d\cos \theta = 2d\cos \theta \tag{2.28.1}$$

where ($\angle OS_2'P = \angle \theta$) and n_0 is the refractive index of the air.

When M_1 is parallel to M_2', d remains the same and δ is only dependent on θ, so the interference fringes should be circular with a point light source.

The conditions for bright and dark fringes of are

$$2d\cos \theta = k\lambda \quad \text{(bright fringes)}$$

and

$$2d\cos \theta = \left(k + \frac{1}{2}\right)\lambda, \quad k = 0, 1, 2, \ldots \quad \text{(dark fringes)} \quad (2.28.2)$$

as can be seen from the above conditions.

a. When d is fixed, $\theta = 0$ corresponds to maximum of δ, which means, the center of circles corresponds to the highest interference order.

b. When we look at a specific interference fringe (i.e. both k and δ are fixed) and increase d, from Eq.(2.28.2), in order to keep δ unchanged, $\cos \theta$ must be reduced, and θ increases. The streaks will be seen gushing out of the center one by one and moving outward. Similarly, when d decreases, the fringes shrink inward and disappear one by one in the center. The appearance or disappearance of a ring corresponds to change in the distance between S_1' and S_2' by a wavelength, λ. And d changes by half a wavelength $\lambda/2$. If M_1 moves by a distance Δd, and the corresponding number of rings is N, then

$$\Delta d = \frac{1}{2}N\lambda \quad (2.28.3)$$

The Δd, wavelength λ can be measured by reading out the interferometer and counting the corresponding ring number change N.

c. It can be proved that the angular spacing of the adjacent two stripes is $\Delta \theta = \dfrac{\lambda}{2d\sin \theta}$, so the smaller the distance d, the greater the stripe spacing and the sparser the stripes. When d is fixed, the smaller θ is, the closer $\Delta \theta$ is to the center, the larger the stripe spacing, otherwise, the denser the stripes.

We can also measure the refractive index of air by using changes in air pressure that cause non-localized interference fringes. An air chamber of length l is placed in the arm of the fixed mirror then pump air into the chamber until the air pressure inside is Δp higher than the ambient atmospheric pressure to obtain stable interference fringes. Slowly release air and count the number of stripes change until the air pressure is balanced and the stripes are stable, record the total number of stripe changes k. Then Δp and Δn meets:

$$2\Delta nl = k\lambda \quad (2.28.4)$$

If the pressure is not too big, the increase of air pressure relative to the that of vacuum p_{amb} and the change of refractive index is proportional to the molecular density, that is, proportional to the pressure, so it can be proportionally calculated how much the refractive index is increased relative to the vacuum under the ambient pressure, thus get the refractive index of air. We already know that Δp in air pressure results in the refractive index changes

Δn, then the refractive index of air under ambient pressure is:

$$n = 1 + \Delta n \cdot \frac{P_{amb}}{\Delta p} = 1 + \frac{k\lambda}{2l} \cdot \frac{P_{amb}}{\Delta p} \quad (2.28.5)$$

②Localized Interference. The actual light source has a certain size (not a geometric point), and is an extended light source. In this experiment, the dispersed laser beam is then scattered by the frosted glass, which can be regarded as an extended light source. Extended light sources can be seen as consisting of numerous unrelated point light sources. When the interferometer is illuminated by the extended light source, these point light sources each produce a set of interference patterns in the space, and these patterns are displaced by a certain amount from each other. Only in a specific area of the overlay field, those interference stripes coincide with each other to form the visible interference stripes. This interference, which can only form interference streaks in a specific area, is called "localized interference".

When the extended light source shines on the Michelson interferometer, equal-inclination or equal-thickness interference occurs depending on the relative position of M_1 and M_2'.

a. Equal-inclination interference. As shown in Fig. 2.28.4, it is assumed that are parallel to each other. Now look at a light ray with an incident angle θ. The two rays reflected by M_1 and M_2' and the optical path difference δ can be calculated as follows:

$$\delta = AB + BC - AD = \frac{2d}{\cos\theta} - 2d \cdot \tan\theta \cdot \sin\theta$$

$$= 2d\left(\frac{1}{\cos\theta} - \frac{\sin^2\theta}{\cos\theta}\right) = 2d\cos\theta \quad (2.28.6)$$

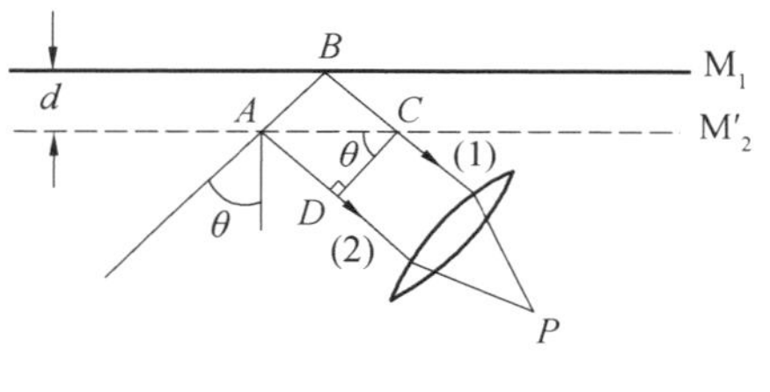

Fig. 2.28.4 The emergence of equal-inclination interference

Here we consider the optical path difference of the reflected rays reflected from parallel mirrors M_1 and M_2'. Because the spacing d of the two mirrors is even, the path difference in formula (2.28.6) is only determined by the incident angle θ, which means that all points that make up the extended light source contribute to the same place in the interference stripe as long as the incident direction is the same. So these fringes are called equal-inclination fringes. Parallel light rays can be thought to intersect at infinite, so the equal-inclination fringes is at infinity. If the reflected light ray is allowed to pass through a converging lens along the normal direction of the mirror, light with an equal reflection angle is focused on a circle on the focal plane. The human eye is equivalent to an auto-focused converging lens, when we look at the equal-inclination fringes, the light of different angle of reflection is focused on circles with

Part 2　Topics of Experiments

different diameters of the retina, and then considering that the different diameters of the circles corresponding to different optical path differences, so we will see the light and dark concentric rings.

b. Equal-thickness interference. When there is a very small angle φ between M_1 and M_2', the interferometer contains a thin wedge-shaped air layer, as shown in Fig. 2.28.5. A light ray is reflected by M_1 and M_2'. And the reflected rays (1) and (2) intersect near the mirror and produce interference. The countless rays of the extended light source, reflected, interfere near the mirror, while in other areas, all light superimposed on each other can only produce uniform light intensity. So near the mirror is the area of this interference fringes.

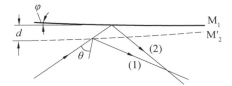

Fig. 2.28.5　Equal-thickness interference

Interference fringes can be imaged by a lens, or by looking directly at the mirror with eyes. We can analyze their properties. When φ is very small, as mentioned earlier, the path difference in light (1) and (2) of light reflected from an incident light is approximately expressed as $\delta = 2d\cos\theta$, where d is the thickness of the air layer near the observation point and θ the angle of incidence. At the intersection of M_1 and M_2', $d = 0$, taking into account the half-wave loss of reflection on the mirror, corresponds to dark fringe, called central line. When the angle θ is small, $\cos\theta \approx 1 - \frac{1}{2}\theta^2$, $\delta \approx 2d\left(1 - \frac{1}{2}\theta^2\right) = 2d - d\theta^2$. Near the central line, the interference stripe is a set of direct lines that are largely parallel to the central line. As the viewing angle increases, to maintain the same difference in path difference, d must be increased, the lines gradually bend, the direction is convex to the central line.

The equal thickness interference of white light: near the intersection of M_1 and M_2' where $d = 0$, interference fringes of white light source can be observed. These stripes are superposition of interference by light with different wavelengths. All but the central line is color stripes symmetric to the center.

3. Content of the experiment

(1) Observe non-localized interference and measure He-Ne laser wavelength.

Turn the beam expander outside the light path and place the screen at O in Fig. 2.28.1. Adjust the He-Ne laser stand so that the beam is parallel to the instrument's top, from the center of the beam splitter to the center of the motion mirror, so that the incident and outlet points of each optical components are about 70 mm in height relative to the table top. Adjust the inclination of the plane mirrors M_1 and M_2 through screws AS_1, AS_2, so that the two sets of light points on the viewing screen coincide, and then place the beam expander in the light path to get the interference fringes on the viewing screen. Further adjust the mirror screw so that the center of the interference rings is in the center of the viewing screen.

To prevent the compensation plate from reflecting light glare, a pinhole screen can be used to mask it.

In rough adjustment, it is normal to see only fine fringes. You can see the change (shrinking in or gushing out) by moving the motion mirror through the knob. The direction of the adjustment is to bring M_1 and closer to each other.

During the adjustment process, determine if M_1 and M_2' are drawing closer, or going away, from each other, when:

①Fringes are born or disappearing from the center;

②The stripes are becoming thicker or thinner;

③The stripes are become denser or sparser.

When the stripes are thick and sparse in the field of view, measure the He-Ne laser wavelength. By moving the motion mirror knob take the reading for each change of 50 stripes repeatedly for 6 times, corresponding to the 0,50,...,250 stripes, and fill in Table 2.28.1 in the report.

(2) Measurement of air refractive index.

Use a low power laser as a light source and place the air chamber with length l of the inner wall in the light path. Adjust the interferometer until non-localized ring interference fringes are observed on the screen. Inflate the chamber (0 - 300 mmHg), then loosen the valve slightly to release the gas slowly. At the same time count the number of stripes going in or coming out k (accurate to half) until the pressure gauge pointer goes back to zero. The refractive index of air is

$$n = 1 + \frac{k\lambda}{2l} \cdot \frac{p_{amb}}{\Delta p} \qquad (2.28.7)$$

where λ is the laser wavelength, and the ambient pressure p_{amb} can be measured by a barometer. Generally take p_{amb} = 760 mmHg. Measure the variation of the stripes under 5 different air pressures and fill in Table 2.28.2 in the experimental report to calculate the corresponding air refractive index.

(3) Observe the localized interference of the extended light source.

In observation of the interference fringes of sodium lamps or other extended light sources, the interferometer can first be adjusted by a laser, and then the laser is replaced with an extended light source to see the interference stripes. You can also adjust the interferometer with a sodium lamp or other light source. Place a pinhole screen on the lamp shade and adjust the two plane mirrors while observing directly to the field of view until the two sets of light points coincide. Remove the pinhole screen and insert a frosted-glass screen between the light source and the beam splitter, and the interference streaks appear.

①Observation of equal-inclination interference fringes. Watch directly with your eyes (focused to infinity). Further adjust the screws and fine-tuning screws on the fixed mirror until the diameter of the stripes remains the same as the eyes move (neither going in nor popping out the center of the circle moves with the eye, and what you see here is the equal-inclination interference fringes.

②Observation of equal-thickness interference fringes. Watch directly with your eyes. On the basis that the equal-inclination interference fringes have been acquired and the stripes become as thick and sparse as possible (at this point early coincide), the fixed mirror is deliberately turned by a little bit, so that M_1 and M_2' slightly deviate from each other until you see parallel straight stripes, which is the equal thickness interference stripe.

(4) Observation of white light interference.

With white light sources (incandescent or white fluorescent lamp), we can see color interference fringes. The fringes are required to be clear and the color is saturated.

Observe and record the characteristics of the color distribution of stripes.

Summarize the conditions for the formation of white light interference and analyze the causes of the color distribution of stripes.

Tips: For M_1 and M_2' to coincide you can first adjust according to non-localized interference condition with a laser point light source until you can see a few thick stripes becoming a straight lines. Replace the laser with an incandescent lamp (or white fluorescent lamp), the color interference stripes may be seen. Slowly adjust the knob of the motion mirror to make the stripes straighter. Because the coherent length of white light is extremely short, when moving the mirror, the operation must be slow and careful, otherwise the fringes will pass by.

Note that the center of the central stripes of all wavelengths in the white light interference is coincident, while the adjacent stripes exhibit distinct colors, and for zero and its adjacent orders, fringes have high visibility or contrast, while for the higher interference orders, the visibility quickly drops. This is also one of the characteristics of white light interference.

4. Discussion

(1) Summarize the characteristics of non-localized interference and localized interference.

(2) Summarize the relationship between the tendency of interference fringes change and the change of relative position of M_1 and M_2' in the adjustment process.

(3) Try to design a detection instrument based on the Michelson interferometer to detect changes in a substance or geometric parameter. Draw the path of light and clarify the design idea.

Physics Laboratory Manual

Experimental Report

Michelson Interferometer

Name: Student ID No.: Score:

Date: Partner: Teacher's signature:

1. Purpose

2. The experiment contents, data recording and processing

(1) Measuring He-Ne laser wavelength (Table 2.28.1).

Table 2.28.1 The positions of the motion mirrors reading for every 50 stripes

Number of stripes	0	50	100	150	200	250
Reading l_i/mm						

Using the difference method, the measured movement of the motion mirrors reading for 50 stripes:

$$\Delta d = \frac{(l_6 - l_3) + (l_5 - l_3) + (l_4 - l_1)}{3} =$$

The actual movement is 1/20 of $\Delta d = \frac{1}{2}N\lambda$ the above reading movement, so the laser wavelength is calculated to be

$$\lambda = \frac{2\Delta d}{20N} =$$

The relative error between the standard wavelength 632.8 nm:

$$E = \frac{632.8 - \lambda}{632.8} \times 100\% =$$

(2) Measurement of air refractive index (Table 2.28.2).

Table 2.28.2 The numbers of stripes for every 50 mmHg

Pressure/mmHg	50	100	150	200	250
Number of stripes/k					

Draw the air pressure- number of stripes graph below.

Draw a straight line in the figure that passes through all the experimental points, that is, a fitted line. Take the two points A and B from a distance on the fitted line, read the coordinates, and calculate the slope (S) of the fitted line.

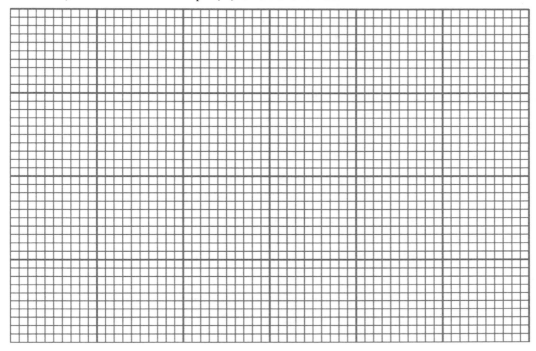

The refractive of air is calculated by

$$n = 1 + \Delta n \cdot \frac{p_{\text{amb}}}{\Delta p} = 1 + \frac{k\lambda}{2l} \cdot \frac{p_{\text{amb}}}{\Delta p} = 1 + \frac{\lambda}{2l} \cdot S \cdot p_{\text{amb}} =$$

3. Discussion

2.29 Franck-Hertz Experiment

In 1913, Niels Bohr predicted that the electrons orbit the nucleus with quantized, discrete energy but can jump from one energy orbit (or level) to another. Basing on Rutherford's nucleus model of the hydrogen, Bohr proposed the model of atom which was the first successful atom model. In 1922, Bohr was awarded the Nobel Prize in Physics "for his services in the investigation of the structure of atoms and of the radiation emanating from them".

In 1914, James Franck and Gustav Hertz published two papers that described their experiments bombarding mercury vapor atoms with slow electrons. In their first paper, Franck and Hertz discovered that the energy exchange between the mercury atoms and the electrons in distinct steps of 4.89 eV through electron-atom collisions. In the second paper, they reported on the light emission at λ = 253.7 nm by the mercury atoms that had absorbed energy from electron-atom collisions. Their experimental results showed that the wavelength of this ultraviolet light corresponded exactly to the 4.89 eV of energy that the flying electron had lost through inelastic collisions. These experiments firstly confirmed the existence of the quantized atomic energy levels that proposed by Bohr. In 1925, Franck and Hertz were awarded the Nobel Prize in Physics for their "discovery of the laws governing the impact of an electron upon an atom".

1. Purpose

(1) Reproduce the Franck-Hertz experiment with Argon atoms, and determine the relationship between the anode current and the accelerating voltage, i.e. the Franck-Hertz curve;

(2) Grasp the experimental method of identifying the quantized energy levels of an atom through measurements of the macroscopic quantities such as the voltage and the current;

(3) Understand the underlying physics of the Franck-Hertz curve;

(4) Understand the physical phenomena of elastic collision and inelastic collision between the electrons and the atoms.

2. Principle

In order to explain the observed line spectra of the elements, in particular the hydrogen spectrum, Bohr predicted that the electron is able to revolve in certain stable orbits around the nucleus without radiating any energy. These stable orbits are called stationary orbits and are attained at certain discrete distances from the nucleus. The electron cannot have any other orbit between these discrete ones. These orbits are associated with definite energies and are also called energy shells or energy levels. Electrons can only gain and lose energy by jumping from one orbit allowed to another, absorbing or emitting electromagnetic radiation with a frequency ν which is determined by the energy difference of the levels according to the Planck relation:

$$\Delta E = E_i - E_j = h\nu \tag{2.29.1}$$

where h is Planck's constant, E_i and $E_j (E_i > E_j)$ denote the energies corresponding to the two energy levels. The energy used to excite an atom from the ground state E_0 to the first excited state E_1 is called the first excitation energy of an atom, which is denoted as $E_0 = E_1 - E_0$ in this experiment.

There are many different ways to excite an atom from low-energy state to high-energy state. For example, an atom can be excited through absorbing certain energies carried by a specific photon, which is called Excited State Absorption (ESA). In Franck-Hertz experiments, the atom is excited through colliding with the flying electrons. There are two kinds of collisions, i.e., elastic collisions and inelastic collisions, which might occur between the argon atoms and the electrons. Collisions play an important role in the Franck-Hertz experiment and will be described thoroughly in the following part.

The classic Franck-Hertz experiment was carried out with mercury. Here we reproduce the Franck-Hertz experiment to study the quantized energy absorption of argon atoms. The apparatus used in the experiment consists of an argon-filled Franck-Hertz tube, a control unit with power supplies and a sensitive DC current amplifier with a shielded cable.

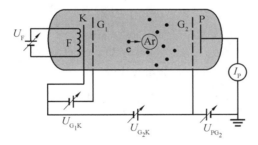

Fig. 2.29.1　Schematic of the four-electrodes Franck-Hertz tube filled with argon vapor and its power supplies used in the experiment

The Franck-Hertz tube is an evacuated glass tube containing low-pressure argon vapor at room temperature. As schematically shown in Fig. 2.29.1, the tube is fitted with the filament F and four electrodes: the cathode K, the perforated control grid G_1 and accelerating grid G_2, and the anode P. The grid G_1 is located in the close proximity to the cathode K. Compared to the distance between K and G_1, the distance between K and G_2 is much longer to guarantee the collisions between the argon atoms and the flying electrons.

The anode P is located adjacent to the perforated grid G_2 to collect the transmitted electrons.

The power supply section of the control unit delivers the filament heating voltage U_F (continuously variable from 0 to 5.0 V), the control voltage U_{G_1K} (continuously variable from 0 to 3.0 V), the accelerating voltage U_{G_2K} (continuously variable from 0 to 85.0 V), and the retarding voltage U_{PG_2} (continuously variable from 0 to 10.0 V). In the experiments, U_{G_1K} is maintained at a small positive voltage to attract more electrons from the filament F and avoid the emitted electrons forming a charge cloud over the surface of the cathode K. U_{G_2K} is a controllable positive voltage to accelerate the flying electrons. U_{PG_2} is a negative voltage to decelerate the electrons trying to reach the anode P. The magnitude of the anode current I_P is of order of 10^{-7} A and measured with a sensitive DC current amplifier as a function of the accelerating voltage U_{G_2K}.

When a certain voltage U_F is applied to the filament, the cathode K is indirectly heated. The electrons, which are near the top of the Fermi distribution in the cathode K, penetrate the potential barrier at the surface and escape from the surface of the cathode K in a process called thermionic emission. The emitted electrons are accelerated toward the anode P passing through the electric field formed mainly by U_{G_2K} since U_{G_1K} is much smaller than U_{G_2K}. After flying through the accelerating grid G_2, the electrons are decelerated by the retarding voltage U_{PG_2}. Only the electrons with energies exceed eU_{PG2} can reach the anode P and give rise to the anode current I_P.

The emitted electrons have a distribution of kinetic energies which is approximately a Maxwell-Boltzmann distribution with a mean energy. For simplicity, here we suppose the electrons escaping from the cathode surface with zero kinetic energy, so they receive the energy ΔE_e when they fly a distance of λ_e away from the cathode surface:

$$\Delta E_e = e \cdot E_{G_2K} = e \cdot E_{G_2K} \cdot \lambda_e / d_{G_2K} \qquad (2.29.2)$$

where e is unit electron charge; E_{G_2K} denotes the electric field intensity exerted on the electron, which is formed by the accelerating voltage U_{G_2K} between K and G_2 with a distance of d_{G_2K}. And λ_e is the mean free path of the electron which is expressed as

$$\lambda_e = \frac{k_B T}{p\sigma} \qquad (2.29.3)$$

where k_B is Boltzmann's constant, T is the tube temperature expressed in Kelvin, p is the pressure of the argon vapor, σ is the cross section for inelastic collisions. If the mean free path of the electron is large enough, the electron may gain enough kinetic energy with the accelerating high enough voltage U_{G_2K}, and the argon atom might be ionized and the Franck-Hertz tube might be destroyed.

When the energy carried by the electron ΔE_e is equal to or larger than the lowest excitation energy of the atom ΔE_0, i.e. $\Delta E_e \geqslant \Delta E_0$, the argon atom can be excited to the first excited state through colliding with the electron and absorbing specific kinetic energy of the electron. The corresponding accelerating voltage is denoted as U_0 for $\Delta E_0 = e U_0$, and U_0 is called the first excitation potential for an atom.

The relationship between the anode current I_P and the accelerating voltage U_{G_2K} is shown in Fig. 2.29.2, which is a typical curvein Franck-Hertz experiment. In fact, the curve presents a current oscillation with periodically recurrent and equidistant maxima and minima.

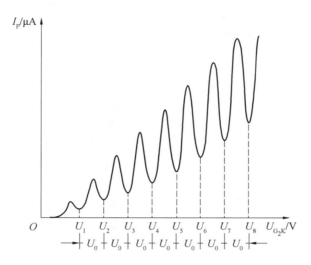

Fig. 2.29.2   Sketch of a typical Franck-Hertz curve

The reason for such a structure of the Franck-Hertz curve is that the electrons undergo inelastic collisions with the argon atoms, resulting in imparting their kinetic energy to the atoms to excite them. As suggested above the anode current I_P is mainly determined by the energies carried by the electrons around the accelerating grid G_2, and the energy of the electrons around G_2 is determined by the value of the accelerating voltage U_{G_2K} and the collisions the electrons experienced between the electrons and the argon atoms.

The electrons flying through the electric field between K and G_2 may collide with the argon atoms. When the energy carried by the electron ΔE_e is smaller than the lowest excitation energy of the atom ΔE_e, i.e. $\Delta E_e < \Delta E_0 (U_{G_2K} < U_0)$, the colliding between the atom and the electron is called elastic collision. The electron hardly loses any energy during elastic collision since the electrons are over one thousand times less than even the lightest atoms. The velocity of the electron remains roughly the same only with its direction may be changed. With the accelerating voltage U_{G_2K} increases from 0 to U_0, more and more electrons with sufficient energies can overcome the retarding voltage U_{PG_2} and therefore the anode current I_P goes up.

When $\Delta E_e \geqslant \Delta E_0 (\text{e.g.}\ U_0 < U_{G_2K} < U_1)$, inelastic collision may occur between the electron and the atom, i.e. the atom can be excited to higher energy level by absorbing a certain amount of the kinetic energy of the electron. The electrons, which participate one inelastic collision just in front of the grid G_2 slows down considerably as well as its direction changed. These electrons are no longer able to reach the anode against the retarding voltage U_{PG_2} and therefore the anode current I_P goes down. Some other electrons still have enough energy to surpass the retarding voltage, and they contribute to the current I_P which never goes back to zero.

When $U_1 < U_{G_2K} < 2U_0$, the electrons participated one inelastic collision continue to be accelerated and the electrons with sufficient energies to contribute to the current I_P is getting more and more. The anode current I_P therefore goes up again after U_{G_2K} exceeds U_1. With the processes similar to the above reappear periodically, multiple inelastic collisions happen in the tube with progressively increasing U_{G_2K} and therefore the cyclic structure of the Franck-Hertz curve is formed.

Assuming the accelerating voltage U_{G_2K} increases linearly with the displacement distance of the flying electrons λ_e, the intervals between adjacent peaks and valleys theoretically corresponds to the first excitation potential of the atom U_0. However, the position of the first anode current maximum is larger than U_0 in the measured Franck-Hertz curves, which is due to the retarding voltage U_{PG_2} and the contact potential between the cathode K and the control grid G_1.

By presenting the relationship between the macroscopic physical quantities of the anode current I_P and the accelerating voltage U_{G_2K}, the Franck-Hertz experiment illustrates the particle-like nature of quantum mechanical processes. The periodicity in the Franck-Hertz curve, i.e. the maxima and the minima are respectively spaced at a fixed accelerating voltage interval, reveals the existence of the discrete, quantized energy states of an atoms.

A common feature of all the excited states is that they "decay" within a very short time, i.e. they "disgorge" their absorbed energy in form of radiation. The decay time is determined by the life time of the excited state, which is usually of the order of about 10^{-8} s. By means of the excitation potential of argon atom, which is about 11.7 eV, one can calculate the wavelength of the emitted photons according to the Einstein relation:

$$\lambda = \frac{hc}{\Delta E_0} = \frac{hc}{eU_0} = \frac{6.63 \times 10^{-34} \times 3.00 \times 10^8}{1.6 \times 10^{-19} \times 11.7} = 1.6 \times 10^2 (\text{nm}) \qquad (2.29.4)$$

It is not easy to observe directly such emission since this is the deep ultraviolet light outside the visible region. Therefore, studying the energy levels of argon atom requires a rather sophisticated experimental equipment.

3. Experimental Contents

During the experiment, we carry out the measurement of the Franck-Hertz curve for argon at room temperature.

(1) Check the connections of all the wires for potentials including U_F, U_{G_1K}, U_{G_2K}, and U_{PG_2}.

(2) Choose the measurement mode as "Manual Mode" and ranges of ammeter as "μA".

(3) Set all the potentials U_F, U_{G_1K}, and U_{PG_2} of as the values noted on the cover of the power supply control unit.

(4) Slowly increase the accelerating voltage U_{G_2K} from 0 to 85.0 V with a step of 0.2 V, and note down the corresponding the anode current I_P on the data sheet.

(5) Choose the measurement mode as "Auto Mode", set the value of U_{G_2K} as 85.0 V, press the "start" button, and the Franck-Hertz curve will be displayed on the oscilloscope working with "X-Y" sweep mode.

(6) Record all the maxima and minima of the Franck-Hertz curve, which is displayed on

the oscilloscope, according to the digital readout of the power supply control unit.

4. Safety Notes

(1) The experimental apparatus needs a warm-up time of 20 – 30 min.

(2) All the values of the potentials cannot exceed the reference values noted on the cover of the control unit.

(3) The value of U_{G_2K} must be one way increased from small to large.

(4) After all the experimental measurements, switch the measurement mode from "Auto" to "Manual" and make sure all the voltages are zero before shut the experimental apparatus down.

5. Discussion

(1) Explain the terms of elastic collision and inelastic collision.

(2) What is the first excitation energy level of an atom?

(3) Why the Franck-Hertz tube used in the experiment just filled with a monatomic gas?

(4) What roles do the control voltage U_{G_1K}, the accelerating voltage U_{G_2K}, and the retarding voltage U_{PG_2} play in the experiment?

Experimental Report

Franck-Hertz Experiment

Name: Student ID No.: Score:

Date: Partner: Teacher's signature:

1. Purpose

2. The experiment contents, data recording and processing

(1) Set the initial conditions for the experiment under the instruction of the teacher and record some important parameters by filling in Table 2.29.1.

Table 2.29.1 The setting parameters of the argon tube

	Typical setting values	Your setting values
U_F/V		
U_{G_1K}/V		
U_{PG_2}/V		

(2) In manual mode, change the U_{G_2K} and write down the change of I_P accordingly by filling in Table 2.29.2 and plot the Franck-Hertz curve on a piece of coordinate paper with the recorded data and identify all the potentials correspond to the maxima and minima in the curve.

(3) Derive from your data the value of the first excitation potential of argon atom with the minima in the curve.

Table 2.29.2 The accelerating voltage U_{G_2K} and the anode current I_P

U_{G_2K}/V	0	0.5	1.0	1.5	2.0	2.5	3.0	3.5	4.0	4.5
$I_P/10^{-7}A$										
U_{G_2K}/V	5.0	5.5	6.0	6.5	7.0	7.5	8.0	8.4	8.6	8.8
$I_P/10^{-7}A$										
U_{G_2K}/V	9.0	9.2	9.4	9.6	9.8	10.0	10.2	10.4	10.6	10.8
$I_P/10^{-7}A$										

Continued Table 2.29.2

U_{G_2K}/V	11.0	11.2	11.4	11.6	11.8	12.0	12.2	12.4	12.6	12.8
$I_P/10^{-7}A$										
U_{G_2K}/V	13.0	13.2	13.4	13.6	13.8	14.0	14.2	14.4	14.6	14.8
$I_P/10^{-7}A$										
U_{G_2K}/V	15.0	15.2	15.4	15.6	15.8	16.0	16.2	16.4	16.6	16.8
$I_P/10^{-7}A$										
U_{G_2K}/V	17.0	17.2	17.4	17.6	17.8	18.0	18.2	18.4	18.6	18.8
$I_P/10^{-7}A$										
U_{G_2K}/V	19.0	19.2	19.4	19.6	19.8	20.0	20.2	20.4	20.6	20.8
$I_P/10^{-7}A$										
U_{G_2K}/V	21.0	21.2	21.4	21.6	21.8	22.0	22.2	22.4	22.6	22.8
$I_P/10^{-7}A$										
U_{G_2K}/V	23.0	23.2	23.4	23.6	23.8	24.0	24.2	24.4	24.6	24.8
$I_P/10^{-7}A$										
U_{G_2K}/V	25.0	25.2	25.4	25.6	25.8	26.0	26.2	26.4	26.6	26.8
$I_P/10^{-7}A$										
U_{G_2K}/V	27.0	27.2	27.4	27.6	27.8	27.0	27.2	27.4	27.6	27.8
$I_P/10^{-7}A$										
U_{G_2K}/V	28.0	28.2	28.4	28.6	28.8	29.0	29.2	29.4	29.6	29.8
$I_P/10^{-7}A$										
U_{G_2K}/V	30.0	30.2	30.4	30.6	30.8	31.0	31.2	31.4	31.6	31.8
$I_P/10^{-7}A$										
U_{G_2K}/V	32.0	32.2	32.4	32.6	32.8	33.0	33.2	33.4	33.6	33.8
$I_P/10^{-7}A$										
U_{G_2K}/V	34.0	34.2	34.4	34.6	34.8	35.0	35.2	35.4	35.6	35.8
$I_P/10^{-7}A$										
U_{G_2K}/V	36.0	36.2	36.4	36.6	36.8	37.0	37.2	37.4	37.6	37.8
$I_P/10^{-7}A$										
U_{G_2K}/V	38.0	38.2	38.4	38.6	38.8	39.0	39.2	39.4	39.6	39.8
$I_P/10^{-7}A$										
U_{G_2K}/V	40.0	40.2	40.4	40.6	40.8	41.0	41.2	41.4	41.6	41.8
$I_P/10^{-7}A$										

Continued Table 2.29.2

U_{G_2K}/V	42.0	42.2	42.4	42.6	42.8	43.0	43.2	43.4	43.6	43.8
$I_P/10^{-7}A$										
U_{G_2K}/V	44.0	44.2	44.4	44.6	44.8	45.0	45.2	45.4	45.6	45.8
$I_P/10^{-7}A$										
U_{G_2K}/V	46.0	46.2	46.4	46.6	46.8	47.0	47.2	47.4	47.6	47.8
$I_P/10^{-7}A$										
U_{G_2K}/V	48.0	48.2	48.4	48.6	48.8	49.0	49.2	49.4	49.6	49.8
$I_P/10^{-7}A$										
U_{G_2K}/V	50.0	50.2	50.4	50.6	50.8	51.0	51.2	51.4	51.6	51.8
$I_P/10^{-7}A$										
U_{G_2K}/V	52.0	52.2	52.4	52.6	52.8	53.0	53.2	53.4	53.6	53.8
$I_P/10^{-7}A$										
U_{G_2K}/V	54.0	54.2	54.4	54.6	54.8	55.0	55.2	55.4	55.6	55.8
$I_P/10^{-7}A$										
U_{G_2K}/V	56.0	56.2	56.4	56.6	56.8	57.0	57.2	57.4	57.6	57.8
$I_P/10^{-7}A$										
U_{G_2K}/V	58.0	58.2	58.4	58.6	58.8	59.0	59.2	59.4	59.6	59.8
$I_P/10^{-7}A$										
U_{G_2K}/V	60.0	60.2	60.4	60.6	60.8	61.0	61.2	61.4	61.6	61.8
$I_P/10^{-7}A$										
U_{G_2K}/V	62.0	62.2	62.4	62.6	62.8	63.0	63.2	63.4	63.6	63.8
$I_P/10^{-7}A$										
U_{G_2K}/V	64.0	64.2	64.4	64.6	64.8	65.0	65.2	65.4	65.6	65.8
$I_P/10^{-7}A$										
U_{G_2K}/V	66.0	66.2	66.4	66.6	66.8	67.0	67.2	67.4	67.6	67.8
$I_P/10^{-7}A$										
U_{G_2K}/V	68.0	68.2	68.4	68.6	68.8	69.0	69.2	69.4	69.6	69.8
$I_P/10^{-7}A$										
U_{G_2K}/V	70.0	70.2	70.4	70.6	70.8	71.0	71.2	71.4	71.6	71.8
$I_P/10^{-7}A$										
U_{G_2K}/V	72.0	72.2	72.4	72.6	72.8	73.0	73.2	73.4	73.6	73.8
$I_P/10^{-7}A$										

Part 2 Topics of Experiments

Continued Table 2.29.2

U_{G_2K}/V	74.0	74.2	74.4	74.6	74.8	75.0	75.2	75.4	75.6	75.8
$I_P/10^{-7}A$										
U_{G_2K}/V	76.0	76.2	76.4	76.6	76.8	77.0	77.2	77.4	77.6	77.8
$I_P/10^{-7}A$										
U_{G_2K}/V	78.0	78.2	78.4	78.6	78.8	79.0	79.2	79.4	79.6	79.8
$I_P/10^{-7}A$										
U_{G_2K}/V	80.0	80.2	80.4	80.6	80.8	81.0	81.2	81.4	81.6	81.8
$I_P/10^{-7}A$										
U_{G_2K}/V	82.0	82.2	82.4	82.6	82.8	83.0	83.2	83.4	83.6	83.8
$I_P/10^{-7}A$										
U_{G_2K}/V	84.0	84.2	84.4	84.6	84.8	85.0				
$I_P/10^{-7}A$										

The graph paper is attached next page.

Find all the maxima and minima of the curve and summarize them in the table you designed below.

(4) Compare the potentials correspond to all the maxima and minima with those measured with the oscilloscope filled in Table 2.29.3.

Table 2.29.3 The maxima and minima positions of the automatically measured Franck-Hertz curve

No.	1	2	3	4	5	6
U_{peak}/V						
U_{valley}/V						

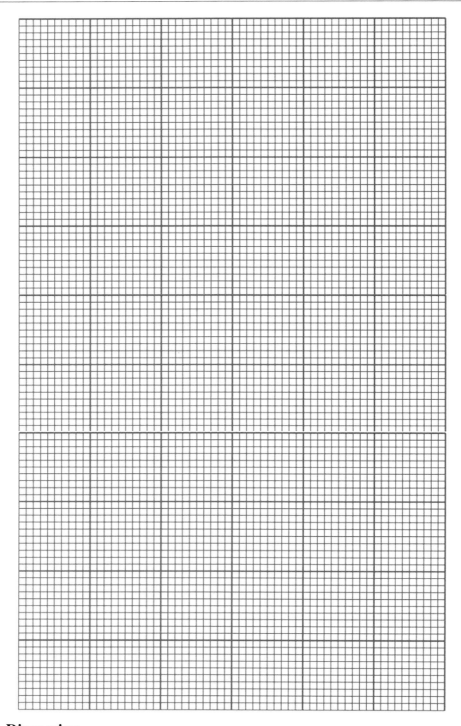

3. Discussion

2.30 Determining the Charge of an Electron by Millikan's Method

1. Purpose

(1) To observe and investigate the characteristics of motion of charged oil droplet in electric and gravitational fields and verify the discreteness of electric charge;

(2) To measure the quantity of electronic charge.

2. Principle and apparatus

The experimental apparatus can be seen in Fig. 2.30.1. Put the atomizer into the oil drop box. Press the atomizer, plenty of oil droplets spray (Fig. 2.30.1(b)) into the box. Part of them may carry electric charges due to friction. If some charged droplets enter the space between two parallel plate electrodes, whose distance is d and voltage we put on it is U, they will experience electric field force and gravity as shown in Fig. 2.30.2. The electric field force is qE with q being quantity of electric charge and $E = U/d$ being electric field intensity. Gravity force is mg with m denoting the mass of an oil droplet and g being the acceleration of gravity. If we choose appropriate voltage U to make these two forces balanced:

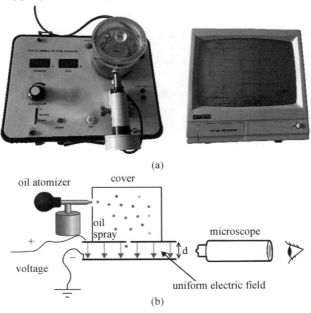

Fig. 2.30.1 (a) Full view of Millikan oil drop apparatus with CCD and screen, the structure of which includes three parts: the oil drop box with two plate electrodes, electric source system (electric source, voltage adjustment and timer) and microscope with CCD detector and display. (b) Schematic view of apparatus of an oil drop box

$$mg = \frac{qU}{d} \qquad (2.30.1)$$

the oil droplet will stand motionless, which is called as balanced state. At this state, the charge quantity of oil droplet can be calculated by measuring the mass of the droplet, distance and voltage between two electrodes.

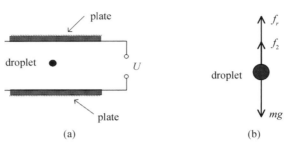

Fig. 2.30.2 (a) The charged droplet between the plate electrodes with voltage difference. (b) The forces that the falling down droplet experiences

It has been proved that the quantity of electric charge is not a successive physical quantity. There is an elementary charge, the electronic charge. Quantity of electric charge of oil droplet must be the integer multiple of elementary charge, namely

$$q = ne \qquad (2.30.2)$$

where n is integer and e represents the value of electronic charge.

In order to find the value of elementary charge, we should measure quantities of electric charges of different oil droplets, q_1, q_2, q_3, ..., and get the least common factor. The least common factor is the elementary charge.

Generally, the mass of oil droplet is about 10^{-15} kg, so it is quite hard to detect it. Oil droplet is too small and its shape nearly is spherical, thus the mass can be reasonably written as

$$m = \frac{4}{3}\pi a^3 \rho \qquad (2.30.3)$$

where ρ and a denote the density of oil and the radius of the oil droplet, respectively. The radius is also hard to measure in fact. However, we can get it by studying the viscosity of oil droplet in air.

For oil droplet that stands motionless between two electrodes, if we set the voltage of electrode zero, it will fall freely due to the gravity. When it falls, it experiences three forces, namely buoyant force by air f_2 and viscous force f_r and gravity mg (Fig. 2.30.2(b)).

Buoyant force can be expressed by $f_2 = 4/3\pi a^3 \rho' g$ according to Archimedes' principle, where ρ' denotes the density of air while according to Stokes' law, viscous force may be written as $f_r = 6\pi a \eta v$ with η being the viscous coefficient and v the falling velocity. It is obvious that viscous force f_r increases with the increasing of droplet's velocity. At the beginning of oil droplet falling, v is small, $f_2 + f_r < mg$ and the oil droplet moves downward rapidly with a downward acceleration. But with the increment of v, f_r increases gradually.

when v reaches a value which makes

$$f_2 + f_r = mg \tag{2.30.4}$$

namely

$$\frac{4}{3}\pi a^3 \rho' g + 6\pi \eta v = \frac{4}{3}\pi a^3 \rho g \tag{2.30.5}$$

the droplet will move uniformly instead of acceleration. Because of $\rho' \ll \rho$, the buoyant force can be omitted. Then

$$6\pi a \eta v = \frac{4}{3}\pi a^3 \rho g \tag{2.30.6}$$

Thus the radius of droplet can be obtained by

$$a = \sqrt{\frac{9\eta v}{2gp}} \tag{2.30.7}$$

Actually, the diameter is too small to treat air as continuum, so the Stokes' law should be revised as

$$f_r = \frac{6\pi a \eta \cdot v}{1 + \dfrac{b}{ap}} \tag{2.30.8}$$

with $b = 0.008\,22$ m $\cdot$ Pa and p being atmospheric pressure with unit Pa. The radius of droplet is rewritten as

$$a = \sqrt{\frac{9\eta v}{2gp\left(1 + \dfrac{b}{ap}\right)}} \tag{2.30.9}$$

Combining Eq.(2.30.1) and Eq.(2.30.2), we get the quantity of electric charge of an oil droplet:

$$q = ne = \frac{18\pi}{\sqrt{2g\rho}} \cdot \frac{d}{U} \left(\frac{\eta v}{1 + \dfrac{b}{ap}} \right)^{\frac{3}{2}} \tag{2.30.10}$$

The velocity of droplet can be measured by measuring the falling time over a given distance, namely

$$v = \frac{l}{t} \tag{2.30.11}$$

where $l = 2$ mm which corresponds to a span from the second horizontal line to penultimate horizontal line of screen. Substituting Eq.(2.30.9) and Eq.(2.30.11) into Eq.(2.30.10), we get the expression of quantity of electric charge:

$$q = ne = \frac{18\pi}{\sqrt{2g\rho}} \cdot \frac{d}{U} \left[\frac{\eta l}{t\left(1 + \dfrac{b}{P}\sqrt{\dfrac{2\rho g t}{9l\eta}}\right)} \right]^{\frac{3}{2}} \tag{2.30.12}$$

The parameters ρ, g, η, P, b, d, l are constants related to experimental conditions and apparatuses:

$$\rho = 981 \text{ kg} \cdot \text{m}^{-3}, \quad g = 9.80 \text{ m} \cdot \text{s}^2$$
$$\eta = 1.83 \times 10^{-5} \text{ kg} \cdot \text{m}^{-1} \cdot \text{s}^{-1}, \quad p = 1.013 \times 10^5 \text{ Pa}$$
$$b = 8.22 \times 10^{-3} \text{ Pa} \cdot \text{m}, \quad d = 5.00 \times 10^{-3} \text{ m}, \quad l = 2.00 \times 10^{-3} \text{ m}$$

Thus Eq.(2.30.12) can be rewritten as

$$q = \frac{1.43 \times 10^{-14}}{[t(1 + 0.02\sqrt{t})]^{\frac{3}{2}}} \cdot \frac{1}{U} \tag{2.30.13}$$

To measure the quantity of an oil droplet, we need measure the balanced voltage first and secondly let it fall freely in air by removing the voltage and record the falling time during a given distance.

3. Experiment contents

(1) Press the oil atomizer and observe movement of droplets by changing the voltage of electrodes.

(2) Choose three oil droplets and record each droplet's voltage needed for balance one time and falling time five times. Fill in the blanks in Table 2.30.1 in the experimental report.

Note: the balance voltage should be in the range of 150 V and 300 V and falling time from the second line to the sixth line on screen should be in the range of 10 s and 30 s (why?).

(3) Calculate the quantity of electric charge for each oil droplet.

(4) Use the known standard value of the elementary charge $e_0 = 1.602 \times 10^{-19}$ C to divide q we have obtained, round up the result and get an integer n at last, i.e.:

$$n \approx \frac{q}{e_0}$$

Then calculate the electronic charge carried by the droplet with

$$e = \frac{q}{n}$$

For each droplet we can get a value of e. Calculate the average of three values of e_1, e_2, e_3, namely

$$\bar{e} = \frac{e_1 + e_2 + e_3}{3}$$

At last, compare $\bar{e}$ and e_0, calculate the absolute difference by $\Delta e = |\bar{e} - e_0|$ and the relative difference by $E = \Delta e / \bar{e} \times 100\%$.

4. Discussion

(1) How does the mass reduction of oil droplet induced by volatilization influence U, m, and q?

(2) Try to explain the lateral movement of an oil droplet.

Part 2 Topics of Experiments

Experimental Report

Determining the Charge of an Electron by Millikan's Method

Name: Student ID No.:   Score:

Date: Partner: Teacher's signature:

1. Purpose

2. The experimental data (Table 2.30.1)

Table 2.30.1 Balance voltage and falling time

Drop No.	Balance voltage (U)	Falling time (t)					Average of t
		1	2	3	4	5	
1							
2							
3							

3. Data processing

(1) Calculate the quantity of electric charge for each oil droplet by Eq. (2.30.8).

$q_1 =$

$q_2 =$

$q_3 =$

(2) Calculate the round-up integer n where $e_0 = 1.602 \times 10^{-19}$ C.

$n_1 \approx \dfrac{q_1}{e_0} =$

269

$$n_2 \approx \frac{q_2}{e_0} =$$

$$n_3 \approx \frac{q_3}{e_0} =$$

(3) Calculate the electronic charge.

$$e_1 = \frac{q_1}{n_1} =$$

$$e_2 = \frac{q_2}{n_2} =$$

$$e_3 = \frac{q_3}{n_3} =$$

$$\bar{e} = \frac{e_1 + e_2 + e_3}{3} =$$

$$\Delta e = |\bar{e} - e_0| =$$

$$E = \frac{\Delta e}{\bar{e}} \times 100\% =$$

4. Discussion

2.31 Optical Fiber Transmission Technique

It is well known that the network of information has already changed our daily life and work style. In fact the fiber technique lays the key foundation for the network, in which the optical fiber communication plays an important role. There are two key techniques, namely the miniaturization of light sources and the technique for low loss fibers, become the milestones in the development of fiber network. In 2000, Z. I. Alferov and H. Kroemer were awarded the Nobel Prize in Physics "for developing semiconductor heterostructures used in high-speed and optoelectronics" in the light source application. In 2009, C. K. Kao was awarded the Nobel Prize in Physics "for groundbreaking achievements concerning the transmission of light in fibers for optical communication".

Currently, the optical fiber communication and optical fiber sensing are the frontier of science and technology. Optical fiber technology has gradually penetrated into various fields. The optical fiber transmission technology involved in this experiment is the basis of optical fiber communication. With the development of science and technology, the application scope of this technology will be more and more extensive.

1. Purpose

(1) To understand the structure of the optical fiber transmission system for the audio signal and the principle for selecting the main components;

(2) To be familiar with the basic characteristics of semiconductor electro-optic/optoelectronic devices and the testing methods of the basic performance;

(3) To master the debugging technique of optical fiber transmission system.

2. Principle

(1) System composition.

The structure and principle of the optical fiber transmission experiment system for audio signal is shown in Fig. 2.31.1. It mainly consists of three parts, namely an optical signal transmitter (composed of a semiconductor light-emitting diode and its modulation and driving circuit), the transmission optical fiber and the optical signal receiver (composed of a photodiode, $I - V$ conversion circuit and a power amplifier circuit).

In this experiment system, a GaAs semiconductor Light Emitting Diode (LED) with central wavelength of a 0.85 μm is used as the light source, and a Silicon Photo Diode (SPD) with peak response wavelength of 0.8 - 0.9 μm is used as the light detector.

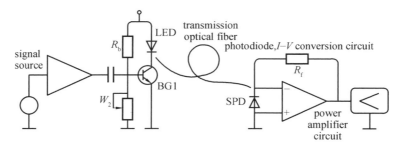

Fig. 2.31.1   General overview of optical fiber communication system

(2) Optical fiber structure and light transmitting principle.

Optical fiber is a dielectric waveguide or medium (such as glass or plastic fiber) through which information (voice, data or video) is transmitted in the form of light. The basic structure of an optical fiber is shown in Fig. 2.31.2. It consists of a transparent core with a refractive index n_1 surrounded by a transparent cladding with a slightly less refractive index n_2. The refractive index of cladding is 1% lower than that of core. For example, the typical values of refractive indices for the core and cladding of a glass fiber are 1.47 and 1.46, respectively. The cladding supports the waveguide structure, protects the core from absorbing surface contaminants and when adequately thick, substantially reduces the radiation loss to the surrounding air. Glass core fibers tend to have low loss in comparison with plastic core fibers. Additionally, most of the fibers are encapsulated in an elastic, abrasion-resistant plastic material which mechanically isolates the fibers from small geometrical irregularities and distortions. A set of guided electromagnetic waves, also called the waveguide modes, can describe the propagation of light along the waveguide. Only a certain number of modes are capable of propagating through the waveguide.

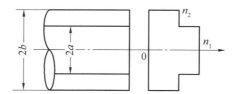

Fig. 2.31.2   The basic structure of an optical fiber with a step-index profile

The optical fiber consists of core and cladding. According to the radial distribution of refractive index along the cross section of optical fibers, optical fibers can be divided into step-index fibers and graded-index fibers. The first one is used in this experiment. The refractive index of the core and cladding are constant, but the refractive index of the core n_1 is slightly larger than that of the cladding. The principle of light transmission can be explained by Total Internal Reflection (TIR) theory, as shown in Fig. 2.31.3. Here, only the incident rays at the end of the optical fiber are considered. The incident plane contains the axis of the optical fiber. According to Snell's law:

$$n_0\sin\theta_i = n_1\sin\theta_z \qquad (2.31.1)$$

$$\theta_z = \frac{\pi}{2} - \alpha \qquad (2.31.2)$$

$$n_0\sin\theta_i = n_1\cos\alpha \qquad (2.31.3)$$

where n_0 is the refractive index of the medium on the left side of the incident plane. When total reflection occurs at the interface between core and cladding, the critical angle of total internal reflection can be expressed as

$$\alpha_c = \arcsin\frac{n_2}{n_1} \qquad (2.31.4)$$

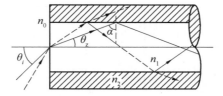

Fig. 2.31.3 The Total Internal Reflection (TIR) in fiber

With the increase of incident angle θ_i, α decreases gradually until $\alpha = \alpha_c$, the meridian ray is confined within the core. Thereafter, if θ_i continues to increase, the angle α will be less than α_c, and some energy will leak out of the core at each reflection. Usually, the end of the optical fiber is in the air, so $n_0 = 1$. When $\alpha = \alpha_c$, the corresponding $\theta_i = \theta_{imax}$

$$n_0\sin\theta_{imax} = n_1\cos\alpha_c \qquad (2.31.5)$$

$$\sin\theta_{imax} = n_1\sqrt{1-\sin^2\alpha_c} = \sqrt{n_1^2 - n_2^2} \qquad (2.31.6)$$

$$NA = \sin\theta_{imax} = \sqrt{n_1^2 - n_2^2} \qquad (2.31.7)$$

Normally, we use numerical aperture NA to characterize the ability of optical fibers to capture meridian rays. From Eq.(2.31.7) one can find that the value of NA is only related to the refractive index of the core n_1 and cladding n_2, but not to the radius of the optical fibers. The larger the NA, the stronger the light-capture ability, that is, the light emitted from the light source is more easily coupled to the core of the optical fiber.

(3) Structure and working principle of semiconductor light emitting diode.

Optical fiber communication system has special requirements for light source devices in many aspects, such as light wavelength, electro-optic efficiency, working lifetime, spectral width and modulation performance. At present, the most commonly used are semiconductor light-emitting diodes and semiconductor lasers. In this experiment, LED is used as light source device.

The semiconductor light-emitting diode often used in optical fiber transmission system is a semiconductor device with N-P-P three-layer structure, as shown in Fig. 2.31.4. When the forward bias is applied to the structure, the N-layer can inject conductive electrons into the active layer. Once these conductive electrons enter the active layer, they can no longer enter the right P layer because of the blocking effect of the right heterojunction. They can only be confined to the active layer and recombined with holes. In the process of recombination, photons with energy are released satisfying the following relations:

$$h\nu = E_1 - E_2 = E_g \qquad (2.31.8)$$

where h is Planck's constant; ν is the frequency of light wave; E_1 is the energy of conducting electron in active layer; E_2 is the energy of conducting electron in valence band state after recombination with hole.

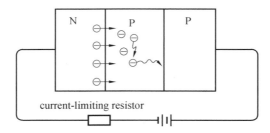

Fig. 2.31.4 Working principle of semiconductor light emitting diode

The light from the semiconductor light-emitting diode used in the optical fiber communication system is output through the pigtail of the optical fiber. The relationship between the light power and the driving current of LED is considered as the electro-optic characteristics of LED. In order to avoid and reduce the non-linear distortion, a proper bias current I should be applied to the LED before use, whose value is equal to the current corresponding to the midpoint of the linear part of the characteristic curve, and the peak value of the modulation signal should be within the linear range of the electro-optic characteristic.

(4) Driving and modulation circuit of LED.

The driving and modulation circuit of the LED at the transmitting end of the audio signal optical fiber transmission system is shown in Fig. 2.31.5. The circuit mainly composed of BG1 is the driving circuit of LED. Adjusting W_2 in this circuit can make the bias current of LED change in the range of 0 – 50 mA. The transmitted audio signal is amplified by an audio amplification circuit mainly composed of IC1, and then coupled to BG1 base via capacitor C_4 to modulate the working current of the LED, so that the LED sends out an optical signal whose light intensity changes with the audio signal, and transmits the signal to the receiving end via the optical fiber.

Part 2 Topics of Experiments

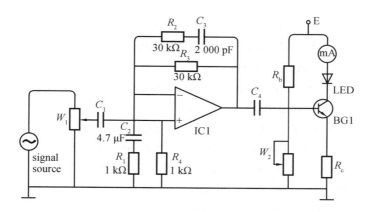

Fig. 2.31.5 Driving and modulation circuit of LED

(5) SPD Optical signal receiver.

Fig. 2.31.6 is a circuit schematic diagram of an optical signal receiver. SPD is a silicon photodiode whose peak response wavelength is close to the center wavelength of the LED light source at the transmitter. Its response to the peak wavelength is 0.25 - 0.5 μA/μW. The task of SPD is to convert the power of the optical signal output from the output end of transmission optical fiber to the optical current I_0 which is proportional to it, and then convert the optical current to the voltage V_0 output through the $I - V$ conversion circuit composed of IC1. The proportion between V_0 and I_0 is

$$V_0 = R_f I_0 \qquad (2.31.9)$$

The IC2 (LA4102) is an integrated audio power amplifier circuit. The resistance elements (including the feedback resistance) of the circuit are integrated in the chip. As long as the external potentiometer W_{nf} is adjusted, the voltage gain of the power amplifier circuit can be changed.

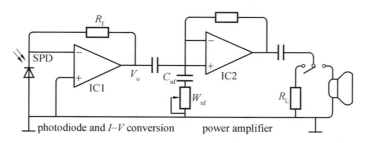

Fig. 2.31.6 Circuit diagram of optical signal receiver

3. Procedures

(1) Modulation and transmission of optical signal.

①The measurement of electro-optic characteristics of LED. The test circuit is shown in Fig. 2.31.7. In addition to the optical power meter, LED and transmission fiber, other components of the circuit are installed in the optical signal transmitter of YOF-B "audio signal optical fiber transmission technology experiment instrument". Before measurement, first insert one end of the cable with current plug at both ends into the current jack on the optical fiber winding coil, and the other end into the "LED" jack on the front panel of the transmitter, and insert the photoelectric probe into the coaxial jack on the optical fiber winding coil, the output end of the transmission optical fiber. The output line of the SPD is connected to the corresponding socket of the optical power meter, then the W_2 is adjusted (corresponding to the "bias adjustment" knob on the front panel of the transmitter), so that the milliammeter is indicated to be any value in the 0-20 mA range, and the reading of the optical power indicator is observed. Under the condition of keeping the LED bias current unchanged, rotate the photoelectric probe around the axis of the coaxial jack appropriately until the reading of the optical power meter is large enough, and pay attention to keep the position of the photoelectric probe unchanged in the future experiment. During measurement, adjust W_2 to make the indication of milliammeter increase gradually from zero. Read the indication of optical power meter every 2 mA until 20 mA. According to the measurement results, the electro-optic characteristic curve of LED transmission optical fiber module is described, and the line segment with good linearity is determined.

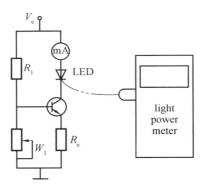

Fig. 2.31.7 Measurement of the electro-optic characteristics of the transmission fiber module LED

②Measurement of the relationship between LED bias current and maximum modulation amplitude (no cut-off distortion). In Fig. 2.31.5, the audio signal generator is used as the signal source, its frequency is adjusted to 1 kHz. One input channel of the dual channel oscilloscope is bridged at both ends of R_e. Then, when the LED bias current is 4 mA, 6 mA, ..., 14 mA, the output amplitude of the signal source is adjusted, and the optical signal with the maximum modulation amplitude without cut-off distortion is observed with an oscilloscope.

③Measurement of amplitude frequency characteristics of modulation and amplification circuit of optical signal transmitter. As shown in Fig. 2.31.5, under the condition of keeping the signal amplitude at the input end of the modulation amplifier unchanged (for example, the peak to peak value is 20 mV), change the signal source frequency within the range of

20 Hz to 15 kHz, and observe the peak to peak value of the waveform at the output end of the amplifier with a dual channel oscilloscope. The amplitude frequency characteristic curve is drawn from the observation results.

(2) Reception of optical signal.

①Measurement of photoelectric characteristics of silicon photodiodes. The test circuit is shown in Fig. 2.31.8. Under the condition of keeping the best coupling state between LED pigtail and SPD, adjust W_2 in LED driving circuit again, so that the bias current of LED increases gradually from zero to 20 mA, read the output voltage V_o(mV) of current voltage converter composed of IC1 once for every 2 mA increase, and draw the $P-V$ transformation curve according to the corresponding optical power P value measured in experiment procedure (1).

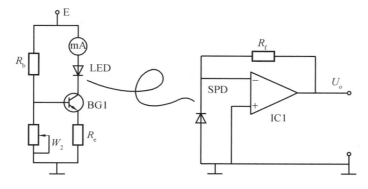

Fig. 2.31.8　Measurement of photoelectric characteristics of photodiodes

②Detection of optical signal. Under the condition of keeping the coupling state of LED tail fiber and receiver photoelectric detection element unchanged and setting the bias current of LED at the transmitter to 10 mA, adjust the amplitude of audio signal source (frequency to 1 kHz) at the transmitter, observe the waveform change of output voltage of $I-V$ conversion circuit at the receiver with oscilloscope, and record the maximum peak to peak value of this waveform without cut-off distortion at the frequency of 1 kHz.

③Transmission of audio signal. Connect the input end of the transmitter to the radio signal, and connect the output end of the receiver power amplifier to the 4 Ω speaker to test the audio effect of the audio signal optical fiber transmission system. During the test, the LED bias current of the transmitter, the input signal amplitude or the resistance value of the potentiometer W_{nf} in the receiver power amplifier circuit can be properly adjusted, the auditory effect of the transmission system can be inspected, and the waveform changes of the input and output signals of the system can be monitored with an oscilloscope.

4. Discussion

(1) When selecting semiconductor light-emitting diodes, how to consider the central wavelength of their light-emitting?

(2) How to set the bias current and modulation amplitude of LED in order to avoid and reduce the nonlinear distortion?

(3) What are the requirements for the peak response wavelength of the selected silicon photodiode (SPD) in this experiment? What is its main task?

Experimental Report

Optical Fiber Transmission Technique

Name:	Student ID No.:	Score:

Date:	Partner:	Teacher's signature:

1. Purpose

2. The experimental data

(1) Determination of $V - I$ characteristics of LED (Table 2.31.1).

Table 2.31.1 Determination of V-I characteristics of LED

V/mV	1 000	1 050	1 100	1 150	1 200	1 250	1 300	1 350	1 400
I/mA									
V/mV	1 450	1 500	1 550	1 600	1 650				
I/mA									

(2) Measurement of electro-optical characteristics of LED (Table 2.31.2).

Table 2.31.2 Measurement of electro-optical characteristics of LED

I/mA	0	5.0	10.0	15.0	20.0	25.0	30.0	35.0	40.0
P/μW									
I/mA	45.0	50.0							
P/μW									

(3) Measurement of optical and electrical characteristics of SPD (Table 2.31.3).

Table 2.31.3 Measurement of optical and electrical characteristics of SPD

P/μW	0	5.0	10.0	15.0	20.0	25.0	30.0	35.0	40.0
V/mV									
P/μW	45.0	50.0							
V/mV									

(4) Measurement of the best working point (bias current) of the instrument (Table 2.31.4).

Table 2.31.4 Measurement of the best working point (bias current) of the instrument

I/mA	10.0	15.0	20.0	25.0	30.0	35.0	40.0	45.0	50.0
$U_{\text{p-p}}$/mV									

3. Data processing

(1) Based on the data in Table 2.31.1, plot a graph of $I - V$ and discuss the characteristics of the curve.

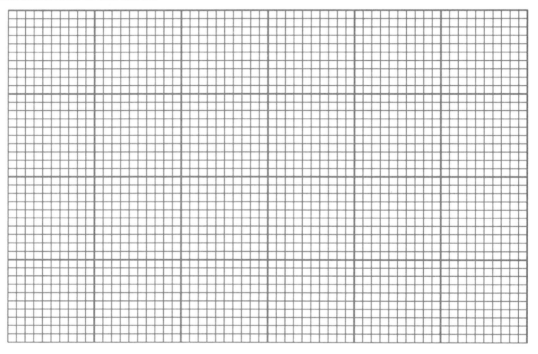

(2) Based on the data in Table 2.31.2, plot a graph of $P - I$ and discuss the characteristics of the curve.

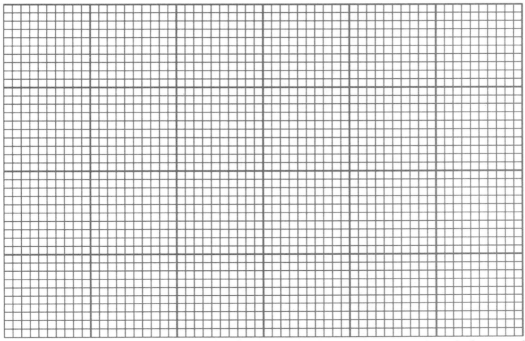

(3) Based on the data in Table 2.31.3, plot a graph of $V - P$ and discuss the characteristics of the curve.

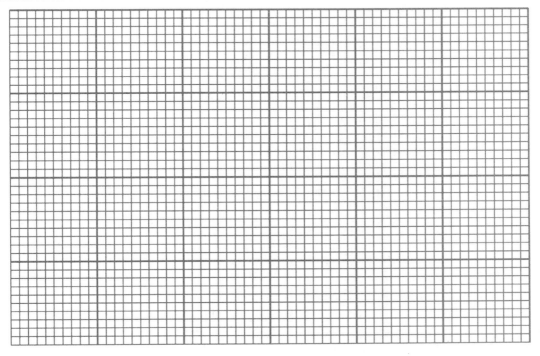

4. Discussion

Part 3　Reading Materials

3.1　The Fundamental Physical Constants

Table 3.1.1　The fundamental physical constants（物理学基本常数表）

Quantity	Symbol or abbreviation	Value	Unit
Speed of light in vacuum	c	$2.997\ 924\ 58 \times 10^8$	$m \cdot s^{-1}$
Gravitational constant	G	$6.674\ 28 \times 10^{-11}$	$m^3 \cdot kg^{-1} \cdot s^{-2}$
Avogadro number	N_A	$6.022\ 141\ 79 \times 10^{23}$	mol^{-1}
Gas constant	R	$8.314\ 472$	$J \cdot mol^{-1} \cdot K^{-1}$
Stefan-Boltzmann constant	k	$5.669\ 620 \times 10^{23}$	$J \cdot s^{-1}$
Electron charge	e	$-1.602\ 176\ 487 \times 10^{-19}$	C
Atomic mass unit	amu	$1.660\ 538\ 782 \times 10^{-27}$	kg
Electron mass	m_e	$9.109\ 382\ 15 \times 10^{-27}$	kg
Electronic ratio	e/m_e	$-1.758\ 820\ 150 \times 10$	$C \cdot kg^{-1}$
Proton mass	m_P	$1.672\ 621\ 637 \times 10^{-27}$	kg
Neutron mass	m_n	$1.674\ 927\ 211 \times 10^{-27}$	kg
Faraday's constant	F	$9.648\ 533\ 99 \times 10^4$	$C \cdot mol^{-1}$
Permittivity of vacuum	ε_0	$8.854\ 187\ 817 \times 10^{-12}$	$F \cdot m^{-1}$
Permeability of vacuum	μ_0	$1.256\ 637\ 061\ 4 \times 10^{-6}$	$H \cdot m^{-1}$
Electron magnetic moment	μ_e	$-9.284\ 763\ 77 \times 10^{-24}$	$J \cdot T^{-1}$
Proton magnetic moment	μ_p	$1.410\ 606\ 662 \times 10^{-26}$	$J \cdot T^{-1}$
Bohr radius	a_0	$0.529\ 177\ 208\ 59 \times 10^{-10}$	m
Bohr magneton	μ_B	$9.274\ 009\ 15 \times 10^{-24}$	$J \cdot T^{-1}$
Nuclear magneton	μ_N	$5.050\ 783\ 24 \times 10^{-27}$	$J \cdot T^{-1}$
Planck's constant	h	$6.626\ 068\ 96 \times 10^{-34}$	$J \cdot s$
Fine structure constant	a	$7.297\ 352\ 537\ 6 \times 10^{-3}$	
Rydberg constant	R_∞	$1.097\ 373\ 156\ 852\ 7 \times 10^7$	m^{-1}
Electron's compton wavelength	$\lambda_c = h/m_e c$	$2.426\ 310\ 217\ 5 \times 10^{-12}$	m
Mass ratio of proton and electron	m_p/m_e	$1\ 836.152\ 672\ 47$	
Boltzmann constant	k	$1.380\ 662 \times 10^{-23}$	K^{-1}

3.2 List of Laureates of Noble Prize in Physics

Table 3.2.1 List of laureates of Noble prize in physics(物理学诺贝尔奖获得者)

Year	Image	Laureate, Country	Rationale
1901		Wilhelm Conrad Röntgen, Germany	"In recognition of the extraordinary services he has rendered by the discovery of the remarkable rays subsequently named after him"
1902		Hendrik Lorentz, Netherlands	"In recognition of the extraordinary service they rendered by their researches into the influence of magnetism upon radiation phenomena"
1902		Pieter Zeeman, Netherlands	
1903		Antoine Henri Becquerel, France	"For his discovery of spontaneous radioactivity"
1903		Pierre Curie, France	"For their joint researches on the radiation phenomena discovered by Professor Henri Becquerel"
1903		Maria Skłodowska – Curie, Poland	
1904		Lord Rayleigh (John William Strutt), UK	"For his investigations of the densities of the most important gases and for his discovery of argon in connection with these studies"
1905		Philipp Eduard Anton von Lenard, Austria-Hungary Germany	"For his work on cathode rays"
1906		Joseph John Thomson, UK	"For his theoretical and experimental investigations on the conduction of electricity by gases"
1907		Albert Abraham Michelson, USA Poland	"For his optical precision instruments and the spectroscopic and metrological investigations carried out with their aid"
1908		Gabriel Lippmann, France	"For his method of reproducing colours photographically based on the phenomenon of interference"

Continued Table 3.2.1

Year	Image	Laureate, Country	Rationale
1909		Guglielmo Marconi, Italy	"For their contributions to the development of wireless telegraphy"
		Karl Ferdinand Braun, Germany	
1910		Johannes Diderik van der Waals, Netherlands	"For his work on the equation of state for gases and liquids"
1911		Wilhelm Wien, Germany	"For his discoveries regarding the laws governing the radiation of heat"
1912		Nils Gustaf Dalén, Sweden	"For his invention of automatic valves designed to be used in combination with gas accumulators in light houses and buoys"
1913		Heike Kamerlingh-Onnes, Netherlands	"For his investigations on the properties of matter at low temperatures which led, inter alia, to the production of liquid helium"
1914		Max von Laue, Germany	"For his discovery of the diffraction of X – rays by crystals", an important step in the development of X – ray spectroscopy
1915		William Henry Bragg, UK	"For their services in the analysis of crystal structure by means of X-rays", an important step in the development of X-ray crystallography
		William Lawrence Bragg, Australia UK	
1916			No Nobel Prize was awarded this year!
1917		Charles Glover Barkla, UK	"For his discovery of the characteristic Röntgen radiation of the elements, another important step in the development of X-ray spectroscopy"
1918		Max Planck, Germany	"For the services he rendered to the advancement of physics by his discovery of energy quanta"

Continued Table 3.2.1

Year	Image	Laureate, Country	Rationale
1919		Johannes Stark, Germany	"For his discovery of the Doppler effect in canal rays and the splitting of spectral lines in electric fields"
1920		Charles Édouard Guillaume, Switzerland	"For the service he has rendered to precision measurements in physics by his discovery of anomalies in nickel-steel alloys"
1921		Albert Einstein, Germany Switzerland	"For his services to theoretical physics, and especially for his discovery of the law of the photoelectric effect"
1922		Niels Bohr, Denmark	"For his services in the investigation of the structure of atoms and of the radiation emanating from them"
1923		Robert Andrews Millikan, USA	"For his work on the elementary charge of electricity and on the photoelectric effect"
1924		Manne Siegbahn, Sweden	"For his discoveries and research in the field of X-ray spectroscopy"
1925		James Franck, Germany	"For their discovery of the laws governing the impact of an electron upon an atom"
		Gustav Hertz, Germany	
1926		Jean Baptiste Perrin, France	"For his work on the discontinuous structure of matter, and especially for his discovery of sedimentation equilibrium"
1927		Arthur Holly Compton, USA	"For his discovery of the effect named after him"
		Charles Thomson Rees Wilson, UK	"For his method of making the paths of electrically charged particles visible by condensation of vapour"
1928		Owen Willans Richardson, UK	"For his work on the thermionic phenomenon and especially for the discovery of the law named after him"

Continued Table 3.2.1

Year	Image	Laureate, Country	Rationale
1929		Louis Victor Pierre Raymond, 7th Duc de Broglie, France	"For his discovery of the wave nature of electrons"
1930		Chandrasekhara Venkata Raman, India	"For his work on the scattering of light and for the discovery of the effect named after him"
1931		No Nobel Prize was awarded this year!	
1932		Werner Heisenberg, Germany	"For the creation of quantum mechanics, the application of which has, inter alia, led to the discovery of the allotropic forms of hydrogen"
1933		Erwin Schrödinger, Austria	"For the discovery of new productive forms of atomic theory"
1933		Paul Dirac, UK	
1934		No Nobel Prize was awarded this year!	
1935		James Chadwick, UK	"For the discovery of the neutron"
1936		Victor Francis Hess, Austria	"For his discovery of cosmic radiation"
1936		Carl David Anderson, USA	"For his discovery of the positron"
1937		Clinton Joseph Davisson, USA	"For their experimental discovery of the diffraction of electrons by crystals"
1937		George Paget Thomson, UK	
1938		Enrico Fermi, Italy	"For his demonstrations of the existence of new radioactive elements produced byneutron irradiation, and for his related discovery of nuclear reactions brought about by slow neutrons"

Continued Table 3.2.1

Year	Image	Laureate, Country	Rationale
1939		Ernest Lawrence, USA	"For the invention and development of the cyclotron and for results obtained with it, especially with regard to artificial radioactive elements"
1940 1941 1942			No Nobel Prize was awarded these years!
1943		Otto Stern, USA Germany	"For his contribution to the development of the molecular ray method and his discovery of the magnetic moment of the proton"
1944		Isidor Isaac Rabi, USA Poland	"For his resonance method for recording the magnetic properties of atomic nuclei"
1945		Wolfgang Pauli, Austria	"For the discovery of the Exclusion Principle, also called the Pauli principle"
1946		Percy Williams Bridgman, USA	"For the invention of an apparatus to produce extremely high pressures, and for the discoveries he made there within the field of high pressure physics"
1947		Edward Victor Appleton, UK	"For his investigations of the physics of the upper atmosphere especially for the discovery of the so-called Appleton layer"
1948		Patrick Maynard Stuart Blackett, UK	"For his development of the Wilson cloud chamber method, and his discoveries there with in the fields of nuclear physics and cosmic radiation"
1949		Hideki Yukawa, Japan	"For his prediction of the existence of mesons on the basis of theoretical work on nuclear forces"
1950		Cecil Frank Powell, UK	"For his development of the photographic method of studying nuclear processes and his discoveries regarding mesons made with this method"
1951		John Douglas Cockcroft, UK Ernest Thomas Sinton Walton, Ireland	"For their pioneer work on the transmutation of atomic nuclei by artificially accelerated atomic particles"

Continued Table 3.2.1

Year	Image	Laureate, Country	Rationale
1952		Felix Bloch, Switzerland USA	"For their development of new methods for nuclear magnetic precision measurements and discoveries in connection therewith"
		Edward Mills Purcell, USA	
1953		Frits Zernike, Netherlands	"For his demonstration of the phase contrast method, especially for his invention of the phase contrast microscope"
1954		Max Born, West Germany	"For his fundamental research in quantum mechanics, especially for his statistical interpretation of the wavefunction"
		Walther Bothe, West Germany	"For the coincidence method and his discoveries made therewith"
1955		Willis Eugene Lamb, USA	"For his discoveries concerning the fine structure of the hydrogen spectrum"
		Polykarp Kusch, USA Germany	"For his precision determination of the magnetic moment of the electron"
1956		John Bardeen, USA	"For their researches on semiconductors and their discovery of the transistor effect"
		Walter Houser Brattain, USA	
		William Bradford Shockley, USA	
1957		Tsung-Dao Lee, China	"For their penetrating investigation of the so-called parity laws which has led to important discoveries regarding the elementary particles"
		Chen-Ning Yang, China	

Continued Table 3.2.1

Year	Laureate, Country	Rationale
1958	Pavel Alekseyevich Cherenkov, Soviet Union	"For the discovery and the interpretation of the Cherenkov effect"
1958	Ilya Frank, Soviet Union	
1958	Igor Yevgenyevich Tamm, Soviet Union	
1959	Emilio Gino Segrè, Italy USA	"For their discovery of the antiproton"
1959	Owen Chamberlain, USA	
1960	Donald Arthur Glaser, USA	"For the invention of the bubble chamber"
1961	Robert Hofstadter, USA	"For his pioneering studies of electron scattering in atomic nuclei and for his thereby achieved discoveries concerning the structure of the nucleons"
1961	Rudolf Ludwig Mössbauer, West Germany	"For his researches concerning the resonance absorption of gamma radiation and his discovery in this connection of the effect which bears his name"
1962	Lev Davidovich Landau, Soviet Union	"For his pioneering theories for condensed matter, especially liquid helium"
1963	Eugene Paul Wigner, Hungary USA	"For his contributions to the theory of the atomic nucleus and the elementary particles, particularly through the discovery and application of fundamental symmetry principles"
1963	Maria Goeppert-Mayer, USA	"For their discoveries concerning nuclear shell structure"
1963	J. Hans D. Jensen, West Germany	

Continued Table 3.2.1

Year	Image	Laureate, Country	Rationale
1964		Nicolay Gennadiyevich Basov, Soviet Union	"For fundamental work in the field of quantum electronics, which has led to the construction of oscillators and amplifiers based on the maser-laser principle"
		Alexander Prokhorov, Soviet Union	
		Charles Hard Townes, USA	
1965		Richard Phillips Feynman, USA	"For their fundamental work inquantum electrodynamics (QED), with deep-ploughing consequences for the physics of elementary particles"
		Julian Schwinger, USA	
		Shin'ichirō Tomonaga, Japan	
1966		Alfred Kastler, France	"For the discovery and development of optical methods for studying Hertzian resonances in atoms"
1967		Hans Albrecht Bethe, USA Germany	"For his contributions to the theory of nuclear reactions, especially his discoveries concerning the energy production in stars"
1968		Luis Walter Alvarez, USA	"For his decisive contributions to elementary particle physics, in particular the discovery of a large number of resonance states, made possible through his development of the technique of using hydrogen bubble chamber and data analysis"
1969		Murray Gell-Mann, USA	"For his contributions and discoveries concerning the classification of elementary particles and their interactions"
1970		Hannes Olof Gösta Alfvén, Sweden	"For fundamental work and discoveries inmagneto-hydrodynamics with fruitful applications in different parts of plasma physics"
		Louis Néel, France	"For fundamental work and discoveries concerning antiferromagnetism and ferrimagnetism which have led to important applications in solid state physics"

Continued Table 3.2.1

Year	Image	Laureate, Country	Rationale
1971		Dennis Gabor, Hungary UK	"For his invention and development of the holographic method"
1972		John Bardeen, USA	"For their jointly developed theory of superconductivity, usually called the BCS-theory"
		Leon Neil Cooper, USA	
		John Robert Schrieffer, USA	
1973		Leo Esaki, Japan	"For their experimental discoveries regarding tunneling phenomena in semiconductors and superconductors, respectively"
		Ivar Giaever, USA Norway	
		Brian David Josephson, UK	"For his theoretical predictions of the properties of a supercurrent through a tunnel barrier, in particular those phenomena which are generally known as the Josephson effect"
1974		Martin Ryle, UK	"For their pioneering research in radio astrophysics: Ryle for his observations and inventions, in particular of the aperture synthesis technique, and Hewish for his decisive role in the discovery of pulsars"
		Antony Hewish, UK	
1975		Aage Bohr, Denmark	"For the discovery of the connection between collective motion and particle motion inatomic nuclei and the development of the theory of the structure of the atomic nucleus based on this connection"
		Ben Roy Mottelson, Denmark	
		Leo James Rainwater, USA	

Continued Table 3.2.1

Year	Image	Laureate, Country	Rationale
1976		Burton Richter, USA	"For their pioneering work in the discovery of a heavy elementary particle of a new kind"
		Samuel Chao Chung Ting, USA	
1977		Philip Warren Anderson, USA	"For their fundamental theoretical investigations of the electronic structure of magnetic and disordered systems"
		Nevill Francis Mott, UK	
		John Hasbrouck Van Vleck, USA	
1978		Pyotr Leonidovich Kapitsa, Soviet Union	"For his basic inventions and discoveries in the area of low-temperature physics"
		Arno Allan Penzias, USA	"For their discovery of cosmic microwave background radiation"
		Robert Woodrow Wilson, USA	"For their discovery of cosmic microwave background radiation"
1979		Sheldon Lee Glashow, USA	"For their contributions to the theory of the unified weak and electromagnetic interaction between elementary particles, including, inter alia, the prediction of the weak neutral current"
		Abdus Salam, Pakistan	
		Steven Weinberg, USA	

Continued Table 3.2.1

Year	Image	Laureate, Country	Rationale
1980		James Watson Cronin, USA	"For the discovery of violations of fundamental symmetry principles in the decay of neutral K-mesons"
		Val Logsdon Fitch, USA	
1981		Nicolaas Bloembergen, Netherlands USA	"For their contribution to the development of laser spectroscopy"
		Arthur Leonard Schawlow, USA	
		Kai Manne Börje Siegbahn, Sweden	"For his contribution to the development of high-resolution electron spectroscopy"
1982		Kenneth G. Wilson, USA	"For his theory for critical phenomena in connection with phase transitions"
1983		Subrahmanyan Chandrasekhar, India USA	"For his theoretical studies of the physical processes of importance to the structure and evolution of the stars"
		William Alfred Fowler, USA	"For his theoretical and experimental studies of the nuclear reactions of importance in the formation of the chemical elements in the universe"
1984		Carlo Rubbia, Italy	"For their decisive contributions to the large project, which led to the discovery of the field particles W and Z, communicators of weak interaction"
		Simon van der Meer, Netherlands	
1985		Klaus von Klitzing, West Germany	"For the discovery of the quantized Hall effect"

Continued Table 3.2.1

Year	Laureate, Country	Rationale
1986	Ernst Ruska, West Germany	"For his fundamental work in electron optics, and for the design of the first electron microscope"
	Gerd Binnig, West Germany	"For their design of the scanning tunneling microscope"
	Heinrich Rohrer, Switzerland	
1987	Johannes Georg Bednorz, West Germany	"For their important break-through in the discovery of superconductivity in ceramic materials"
	Karl Alexander Müller, Switzerland	
1988	Leon Max Lederman, USA	"For the neutrino beam method and the demonstration of the doublet structure of the leptons through the discovery of the muon neutrino"
	Melvin Schwartz, USA	
	Jack Steinberger, USA	
1989	Norman Foster Ramsey, USA	"For the invention of the separated oscillatory fields method and its use in the hydrogen maser and other atomic clocks"
	Hans Georg Dehmelt, USA Germany	"For the development of the ion trap technique"
	Wolfgang Paul, West Germany	

Continued Table 3.2.1

Year	Image	Laureate, Country	Rationale
1990		Jerome I. Friedman, USA	"For their pioneering investigations concerning deep inelastic scattering of electrons on protons and bound neutrons, which have been of essential importance for the development of the quark model in particle physics"
		Henry Way Kendall, USA	
		Richard E. Taylor, Canada	
1991		Pierre-Gilles de Gennes, France	"For discovering that methods developed for studying order phenomena in simple systems can be generalized to more complex forms of matter, in particular to liquid crystals and polymers"
1992		Georges Charpak, France Poland	"For his invention and development of particle detectors, in particular the multiwire proportional chamber"
1993		Russell Alan Hulse, USA	"For the discovery of a new type of pulsar, a discovery that has opened up new possibilities for the study of gravitation"
		Joseph Hooton Taylor Jr., USA	
1994		Bertram Brockhouse, Canada	"For the development of neutron spectroscopy" and "For pioneering contributions to the development of neutron scattering techniques for studies of condensed matter"
		Clifford Glenwood Shull, USA	"For the development of theneutron diffraction technique" and "For pioneering contributions to the development of neutron scattering techniques for studies of condensed matter"
1995		Martin Lewis Perl, USA	"For the discovery of the tau lepton" and "For pioneering experimental contributions to lepton physics"
		Frederick Reines, USA	

Continued Table 3.2.1

Year	Image	Laureate, Country	Rationale
1996		David Morris Lee, USA	"For their discovery of superfluidity in helium-3"
		Douglas D. Osheroff, USA	
		Robert Coleman Richardson, USA	
1997		Steven Chu, USA	"For development of methods to cool and trap atoms with laser light"
		Claude Cohen-Tannoudji, France	
		William Daniel Phillips, USA	
1998		Robert B. Laughlin, USA	"For their discovery of a new form of quantum fluid with fractionally charged excitations"
		Horst Ludwig Störmer, Germany	
		Daniel Chee Tsui, China USA	
1999		Gerard 't Hooft, Netherlands	"For elucidating the quantum structure of electroweak interactions in physics"
		Martinus J. G. Veltman, Netherlands	

Continued Table 3.2.1

Year	Image	Laureate, Country	Rationale
2000		Zhores Ivanovich Alferov, Russia	"For developing semiconductor heterostructures used in high-speed and optoelectronics"
		Herbert Kroemer, Germany	
		Jack St. Clair Kilby, USA	"For his part in the invention of the integrated circuit"
2001		Eric Allin Cornell, USA	"For the achievement of Bose-Einstein condensation in dilute gases of alkali atoms, and for early fundamental studies of the properties of the condensates"
		Carl Edwin Wieman, USA	
		Wolfgang Ketterle, Germany	
2002		Raymond Davis Jr., USA	"For pioneering contributions to astrophysics, in particular for the detection of cosmic neutrinos"
		Masatoshi Koshiba, Japan	
		Riccardo Giacconi, Italy USA	"For pioneering contributions to astrophysics, which have led to the discovery of cosmic X-ray sources"
2003		Alexei Alexeyevich Abrikosov, Russia USA	"For pioneering contributions to the theory of superconductors and superfluids"
		Vitaly Lazarevich Ginzburg, Russia	
		Anthony James Leggett, UK USA	

Continued Table 3.2.1

Year	Image	Laureate, Country	Rationale
2004		David J. Gross, USA	"For the discovery of asymptotic freedom in the theory of the strong interaction"
		Hugh David Politzer, USA	
		Frank Wilczek, USA	
2005		Roy J. Glauber, USA	"For his contribution to the quantum theory of optical coherence"
		John L. Hall, USA	"For their contributions to the development of laser-based precisionspectroscopy, including the optical frequency comb technique"
		Theodor W. Hänsch, Germany	
2006		John C. Mather, USA	"For their discovery of the blackbody form and anisotropy of the cosmic microwave background radiation"
		George F. Smoot, USA	
2007		Albert Fert, France	"For the discovery of giant magnetoresistance"
		Peter Grünberg, Germany	

Continued Table 3.2.1

Year	Image	Laureate, Country	Rationale
2008		Makoto Kobayashi, Japan	"For the discovery of the origin of the broken symmetry which predicts the existence of at least three families of quarks in nature"
		Toshihide Maskawa, Japan	
		Yoichiro Nambu, Japan USA	"For the discovery of the mechanism of spontaneous broken symmetry in subatomic physics"
2009		Charles K. Kao, Hong Kong UK USA	"For groundbreaking achievements concerning the transmission of light in fibers for optical communication"
		Willard S. Boyle, Canada USA	"For the invention of an imaging semiconductor circuit-the CCD sensor"
		George E. Smith, USA	
2010		Andre Geim, Russia UK Netherlands	"For groundbreaking experiments regarding the two-dimensional material graphene"
		Konstantin Novoselov, Russia UK	
2011		Saul Perlmutter, USA	"For the discovery of the accelerating expansion of the Universe through observations of distant supernovae"
		Brian P. Schmidt, Australia USA	
		Adam G. Riess, USA	

Continued Table 3.2.1

Year	Image	Laureate, Country	Rationale
2012		Serge Haroche, France	"For ground-breaking experimental methods that enable measuring and manipulation of individual quantum systems"
		David J. Wineland, USA	
2013		François Englert, Belgium	"For the theoretical discovery of a mechanism that contributes to our understanding of the origin of mass of subatomic particles, and which recently was confirmed through the discovery of the predicted fundamental particle, by the ATLAS and CMS experiments at CERN's Large Hadron Collider"
		Peter Higgs, UK	
2014		Isamu Akasaki, Japan	"For the invention of efficient blue light-emitting diodes which has enabled bright and energy-saving white light sources"
		Hiroshi Amano, Japan	
		Shuji Nakamura, Japan USA	
2015		Takaaki Kajita, Japan	"For the discovery of neutrino oscillations, which shows that neutrinos have mass"
		Arthur B. McDonald, Canada	
2016		David J. Thouless, UK	"For theoretical discoveries of topological phase transitions and topological phases of matter"
		F. Duncan M. Haldane, UK	
		John M. Kosterlitz, UK USA	

301

Continued Table 3.2.1

Year	Image	Laureate, Country	Rationale
2017		Rainer Weiss, Germany USA	"For decisive contributions to the LIGO detector and the observation of gravitational waves"
		Kip Thorne, USA	
		Barry Barish, USA	
2018		Arthur Ashkin, USA	"For groundbreaking inventions in the field of laser physics", in particular "For the optical tweezers and their application to biological systems"
		Gérard Mourou, France	"For groundbreaking inventions in the field of laser physics", in particular "For their method of generating high-intensity, ultra-short optical pulses"
		Donna Strickland, Canada	
2019		James Peebles, Canada	"For contributions to our understanding of the evolution of the universe and Earth's place in the cosmos"
		Michel Mayor, Switzerland	"For the discovery of an exoplanet orbiting a solar-type star"
		Didier Queloz, Switzerland	
2020		Roger Penrose, UK	"For the discovery that black hole formation is a robust prediction of the general theory of relativity"
		Reinhard Genzel, Germany	"For the discovery of a supermassive compact object at the centre of our galaxy"
		AndreaGhez, USA	

3.3 List of 10 Most Beautiful Experiments in Physics

List of 10 most beautiful experiments in physics:
(1) Young's double-slit experiment applied to the interference of single electrons;
(2) Galileo's experiment on falling bodies (1600s);
(3) Millikan's oil-drop experiment (1910s);
(4) Newton's decomposition of sunlight with a prism (1665 – 1666);
(5) Young's light-interference experiment (1801);
(6) Cavendish's torsion-bar experiment (1798);
(7) Eratosthenes' measurement of the Earth's circumference (3rd century B.C.);
(8) Galileo's experiments with rolling balls down inclined planes (1600s);
(9) Rutherford's discovery of the nucleus (1911);
(10) Foucault's pendulum (1851).

The list above shows the top 10 most frequently mentioned experiments by readers of *Physics World*, which were first commented by Robert P. Crease in his paper entitled with *The most beautiful experiment* (*Physics World* 2002,15(9):19) and described in details in the article entitled with *Here They Are*, *Science's 10 Most Beautiful Experiments* by George Johnson (*New York Time*, 2002,9(24)).

Other experiments concerned by readers of *Physics World*, included:
(11) Archimedes' experiment on hydrostatics;
(12) Roemer's observations of the speed of light;
(13) Joule's paddle-wheel heat experiments;
(14) Reynolds's pipe flow experiment;
(15) Mach & Salcher's acoustic shock wave;
(16) Michelson-Morley measurement of the null effect of the ether;
(17) Röntgen's detection of Maxwell's displacement current;
(18) Oersted's discovery of electromagnetism;
(19) The Braggs' X-ray diffraction of salt crystals;
(20) Eddington's measurement of the bending of starlight;
(21) Stern-Gerlach demonstration of space quantization;
(22) Schrödinger's cat thought experiment;
(23) Trinity test of nuclear chain reaction;
(24) Wu et al.'s measurement of parity violation;
(25) Goldhaber's study of neutrino helicity;
(26) Feynman dipping an O-ring in water.

3.4 More about 10 Most Beautiful Physics Experiments

1. The article from *New York Time*

Here They Are, Science's 10 Most Beautiful Experiments

By George Johnson

Sept. 24, 2002

Whether they are blasting apart subatomic particles in accelerators, sequencing the genome or analyzing the wobble of a distant star, the experiments that grab the world's attention often cost millions of dollars to execute and produce torrents of data to be processed over months by supercomputers. Some research groups have grown to the size of small companies.

But ultimately science comes down to the individual mind grappling with something mysterious. When Robert P. Crease, a member of the philosophy department at the State University of New York at Stony Brook and the historian at Brookhaven National Laboratory, recently asked physicists to nominate the most beautiful experiment of all time, the 10 winners were largely solo performances, involving at most a few assistants. Most of the experiments — which are listed in this month's *Physics World* — took place on table tops and none required more computational power than that of a slide rule or calculator.

What they have in common is that they epitomize the elusive quality scientists call beauty. This is beauty in the classical sense: the logical simplicity of the apparatus, like the logical simplicity of the analysis, seems as inevitable and pure as the lines of a Greek monument. Confusion and ambiguity are momentarily swept aside, and something new about nature becomes clear.

The list in *Physics World* was ranked according to popularity, first place going to an experiment that vividly demonstrated the quantum nature of the physical world. But science is a cumulative enterprise — that is part of its beauty. Rearranged chronologically and annotated below, the winners provide a bird's-eye view of more than 2 000 years of discovery.

(1) Eratosthenes' measurement of the Earth's circumference.

At noon on the summer solstice in the Egyptian town(Syene) now called Aswan, the sun hovers straight overhead: objects cast no shadow and sunlight falls directly down a deep well. When he read this fact, Eratosthenes, the librarian at Alexandria in the third century B.C., realized he had the information he needed to estimate the circumference of the planet. On the same day and time, he measured shadows in Alexandria, finding that the solar rays there had a bit of a slant, deviating from the vertical by about seven degrees.

The rest was just geometry. Assuming the earth is spherical, its circumference spans 360 degrees. So if the two cities are seven degrees apart, that would constitute seven-360ths of the full circle — about one-fiftieth. Estimating from travel time that the towns were 5 000 "stadia" apart, Eratosthenes concluded that the earth must be 50 times that size — 250 000 stadia in girth. Scholars differ over the length of a Greek stadium, so it is impossible to know just how accurate he was. But by some reckonings, he was off by only about 5 percent. (Ranking: 7)

(2) Galileo's experiment on falling objects.

In the late 1500's, everyone knew that heavy objects fall faster than lighter ones. After all, Aristotle had said so, that an ancient Greek scholar still held such sway was a sign of how far science had declined during the dark ages.

Galileo Galilei, who held a chair in mathematics at the University of Pisa, was impudent enough to question the common knowledge. The story has become part of the folklore of science: he is reputed to have dropped two different weights from the town's Leaning Tower showing that they landed at the same time. His challenges to Aristotle may have cost Galileo his job, but he had demonstrated the importance of taking nature, not human authority, as the final arbiter in matters of science. (Ranking: 2)

(3) Galileo's experiments with rolling balls down inclined planes.

Galileo continued to refine his ideas about objects in motion. He took a board 12 cubits long and half a cubit wide (about 20 feet by 10 inches) and cut a groove, as straight and smooth as possible, down the center. He inclined the plane and rolled brass balls down it, timing their descent with a water clock — a large vessel that emptied through a thin tube into a glass. After each run he would weigh the water that had flowed out — his measurement of elapsed time — and compare it with the distance the ball had traveled.

Aristotle would have predicted that the velocity of a rolling ball was constant: double its time in transit and you would double the distance it traversed. Galileo was able to show that the distance is actually proportional to the square of the time: Double it and the ball would go four times as far. The reason is that it is being constantly accelerated by gravity. (Ranking: 8)

(4) Newton's decomposition of sunlight with a prism.

Isaac Newton was born the year Galileo died. He graduated from Trinity College, Cambridge, in 1665, then holed up at home for a couple of years waiting out the plague. He had no trouble keeping himself occupied.

The common wisdom held that white light is the purest form (Aristotle again) and that colored light must therefore have been altered somehow. To test this hypothesis, Newton shined a beam of sunlight through a glass prism and showed that it decomposed into a spectrum cast on the wall. People already knew about rainbows, of course, but they were considered to be little more than pretty aberrations. Actually, Newton concluded, it was these colors — red, orange, yellow, green, blue, indigo, violet and the gradations in between — that were fundamental. What seemed simple on the surface, a beam of white light, was, if one looked deeper, beautifully complex. (Ranking: 4)

(5) Cavendish's torsion-bar experiment.

Another of Newton's contributions was his theory of gravity, which holds that the strength of attraction between two objects increases with the square (the correct word is "product") of their masses and decreases with the square of the distance between them. But how strong is gravity in the first place?

In the late 1700's an English scientist, Henry Cavendish, decided to find out. He took a six-foot wooden rod and attached small metal spheres to each end, like a dumbbell, then suspended it from a wire. Two 350-pound lead spheres placed nearby exerted just enough gravitational force to tug at the smaller balls, causing the dumbbell to move and the wire to twist. By mounting finely etched pieces of ivory on the end of each arm and in the sides of the case, he could measure the subtle displacement. To guard against the influence of air currents, the apparatus (called a torsion balance) was enclosed in a room and observed with telescopes mounted on each side.

The result was a remarkably accurate estimate of a parameter called the gravitational constant, and from that Cavendish was able to calculate the density and mass of the earth. Erastothenes had measured how far around the planet was. Cavendish had weighed it: 6.0×10^{24} kg, or about 13 trillion trillion pounds. (Ranking: 6)

(6) Young's light-interference experiment.

Newton wasn't always right. Through various arguments, he had moved the scientific mainstream toward the conviction that light consists exclusively of particles rather than waves. In 1803, Thomas Young, an English physician and physicist, put the idea to a test. He cut a hole in a window shutter, covered it with a thick piece of paper punctured with a tiny pinhole and used a mirror to divert the thin beam that came shining through. Then he took "a slip of a card, about one-thirtieth of an inch in breadth" and held it edgewise in the path of the beam, dividing it in two. The result was a shadow of alternating light and dark bands — a phenomenon that could be explained if the two beams were interacting like waves.

Bright bands appeared where two crests overlapped, reinforcing each other; dark bands marked where a crest lined up with a trough, neutralizing each other.

The demonstration was often repeated over the years using a card with two holes to divide the beam. These so-called double-slit experiments became the standard for determining wavelike motion — a fact that was to become especially important a century later when quantum theory began. (Ranking: 5)

(7) Foucault's pendulum.

Recent year when scientists mounted a pendulum above the South Pole and watched it swing, they were replicating a celebrated demonstration performed in Paris in 1851. Using a steel wire 220 feet long, the French scientist Jean-Bernard-Léon Foucault suspended a 62-pound iron ball from the dome of the Panthéon and set it in motion, rocking back and forth. To mark its progress he attached a stylus to the ball and placed a ring of damp sand on the floor below.

The audience watched in awe as the pendulum inexplicably appeared to rotate, leaving a slightly different trace with each swing. Actually it was the floor of the Panthéon that was slowly moving, and Foucault had shown, more convincingly than ever, that the earth revolves on its axis. At the latitude of Paris, the pendulum's path would complete a full clockwise rotation every 30 h; on the Southern Hemisphere it would rotate counterclockwise, and on the Equator it wouldn't revolve at all. At the South Pole, as the modern-day scientists confirmed, the period of rotation is 24 h. (Ranking: 10)

(8) Millikan's oil-drop experiment.

Since ancient times, scientists had studied electricity — an intangible essence that came from the sky as lightning or could be produced simply by running a brush through your hair. In 1897 (in an experiment that could easily have made this list), the British physicist J. J. Thomson had established that electricity consisted of negatively charged particles — electrons. It was left to the American scientist Robert Millikan, in 1909, to measure their charge.

Using a perfume atomizer, he sprayed tiny drops of oil into a transparent chamber. At the top and bottom were metal plates hooked to a battery, making one positive and the other negative. Since each droplet picked up a slight charge of static electricity as it traveled through the air, the speed of its descent could be controlled by altering the voltage on the plates. (When this electrical force matched the force of gravity, a droplet — "like a brilliant star on a black background" — would hover in midair.)

Millikan observed one drop after another, varying the voltage and noting the effect. After many repetitions he concluded that charge could only assume certain fixed values. The smallest of these portions was none other than the charge of a single electron. (Ranking: 3)

(9) Rutherford's discovery of the nucleus.

When Ernest Rutherford was experimenting with radioactivity at the University of Manchester in 1911, atoms were generally believed to consist of large mushy blobs of positive electrical charge with electrons embedded inside — the "plum pudding" model. But when he and his assistants fired tiny positively charged projectiles, called alpha particles, at a thin foil of gold, they were surprised that a tiny percentage of them came bouncing back. It was as though bullets had ricocheted off Jell-O.

Rutherford calculated that actually atoms were not so mushy after all. Most of the mass must be concentrated in a tiny core, now called the nucleus, with the electrons hovering around it. With amendments from quantum theory, this image of the atom persists today. (Ranking: 9)

(10) Young's double-slit experiment applied to the interference of single electrons.

Neither Newton nor Young was quite right about the nature of light. Though it is not simply made of particles, neither can it be described purely as a wave. In the first five years of the 20th century, Max Planck and then Albert Einstein showed, respectively, that light is emitted and absorbed in packets — called photons. But other experiments continued to verify that light is also wavelike.

It took quantum theory, developed over the next few decades, to reconcile how both ideas could be true: photons and other subatomic particles — electrons, protons, and so forth — exhibit two complementary qualities; they are, as one physicist put it, "wavicles".

To explain the idea, to others and themselves, physicists often used a thought experiment, in which Young's double-slit demonstration is repeated with a beam of electrons instead of light. Obeying the laws of quantum mechanics, the stream of particles would split in two, and the smaller streams would interfere with each other, leaving the same kind of light and dark striped pattern as was cast by light. Particles would act like waves.

According to an accompanying article in *Physics Today*, by the magazine's editor, Peter Rodgers, it wasn't until 1961 that someone (Claus Jönsson of Tübingen) carried out the experiment in the real world.

By that time no one was really surprised by the outcome, and the report, like most, was absorbed anonymously into science. (Ranking: 1)

2.More detailed information on the experiments excerpted from articles in Wikipedia

(1)On Galileo's experiment on falling bodies.

Between 1589 – 1592, the Italian scientist Galileo Galilei (then professor of mathematics at the University of Pisa) is said to have dropped two spheres of different masses from the Leaning Tower of Pisa(Fig. 3.4.1) to demonstrate that their time of descent was independent of their mass, according to a biography by Galileo's pupil Vincenzo Viviani, composed in 1654 and published in 1717.

Fig. 3.4.1   The Leaning Tower of Pisa, where the experiment supposedly took place

According to the story, Galileo discovered through this experiment that the objects fell with the same acceleration, proving his prediction true, while at the same time disproving Aristotle's theory of gravity (which states that objects fall at speed proportional to their mass). Most historians consider it to have been a thought experiment rather than a physical test as there is little evidence that it actually took place.

At the time when Viviani asserts that the experiment took place, Galileo had not yet formulated the final version of his law of free fall. He had, however, formulated an earlier version which predicted that bodies <u>of the same material</u> falling through the same medium would fall at the same speed. This was contrary to what Aristotle had taught: that heavy objects fall faster than lighter ones, in direct proportion to their weight. While this story has been retold in popular accounts, there is no account by Galileo himself of such an experiment, and it is accepted by most historians that it was a thought experiment which did not actually take place. An exception is Stillman Drake, who argues that it took place, more or less as Viviani described it, as a demonstration for students.

Galileo set out his ideas about falling bodies, and about projectiles in general, in his book *Two New Sciences*. The two sciences were the science of motion, which became the foundation — stone of physics, and the science of materials and construction, an important contribution to engineering. Galileo arrived at his hypothesis by a famous thought experiment outlined in his book *On Motion*. This experiment runs as follows: Imagine two objects, one light and one heavier than the other one, are connected to each other by a string. Drop this system of objects from the top of a tower. If we assume heavier objects do indeed fall faster than lighter ones (and conversely, lighter objects fall slower), the string will soon pull taut as the lighter object retards the fall of the heavier object. But the system considered as a whole is <u>heavier</u> than the heavy object alone, and therefore should fall faster. This contradiction leads one to conclude the assumption is false.

(2) **On Millikan's oil-drop experiment.**

The oil drop experiment (Fig. 3.4.2) was performed by Robert A. Millikan and Harvey Fletcher in 1909 to measure the elementary electric charge (the charge of the electron). The experiment took place in the Ryerson Physical Laboratory at the University of Chicago. Millikan received the Nobel Prize in Physics in 1923.

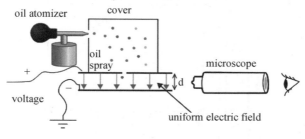

Fig. 3.4.2　Simplified scheme of Millikan's oil drop experiment

The experiment entailed observing tiny electrically charged droplets of oil located between two parallel metal surfaces, forming the plates of a capacitor. The plates were oriented horizontally, with one plate above the other. A mist of atomized oil drops was introduced through a small hole in the top plate and was ionized by an X-ray, making them negatively charged (Fig. 3.4.3). First, with zero applied electric field, the velocity of a falling droplet was measured. At terminal velocity, the drag force equals the gravitational force. As both forces depend on the radius in different ways, the radius of the droplet, and therefore the mass and gravitational force, could be determined (using the known density of the oil). Next, a voltage inducing an electric field was applied between the plates and adjusted until the drops were suspended in mechanical equilibrium, indicating that the electrical force and the gravitational force were in balance. Using the known electric field, Millikan and Fletcher could determine the charge on the oil droplet. By repeating the experiment for many droplets, they confirmed that the charges were all small integer multiples of a certain base value, which was found to be $1.592\ 4(17) \times 10^{-19}$ C, about 0.6% difference from the currently accepted value of $1.602\ 176\ 634 \times 10^{-19}$ C, which was measured on May 2019 (before that the most recent (2014) accepted value was $1.602\ 176\ 620\ 8(98) \times 10^{-19}$ C, where the indicates the uncertainty of the last two decimal places). The difference is less than one percent, but is six times greater than Millikan's standard error, so the disagreement is significant.

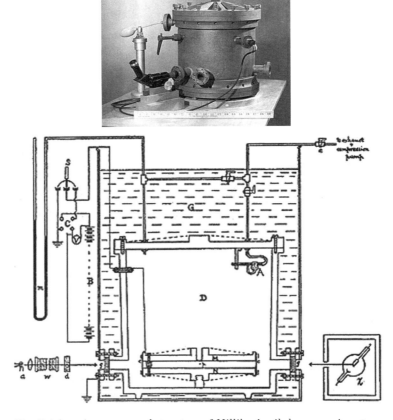

Fig. 3.4.3　Apparatus and structure of Millikan's oil drop experiment

①Detailed descripton of experiment. Millikan's and Fletcher's apparatus incorporated a parallel pair of horizontal metal plates. By applying a potential difference across the plates, a uniform electric field was created in the space between them. A ring of insulating material was used to hold the plates apart. Four holes were cut into the ring, three for illumination by a bright light, and another to allow viewing through a microscope.

A fine mist of oil droplets was sprayed into a chamber above the plates. The oil was of a type usually used in vacuum apparatus and was chosen because it had an extremely low vapour pressure. Ordinary oil would evaporate under the heat of the light source causing the mass of the oil drop to change over the course of the experiment. Some oil drops became electrically charged through friction with the nozzle as they were sprayed. Alternatively, charging could be brought about by including an ionizing radiation source (such as an X-ray tube). The droplets entered the space between the plates and, because they were charged, could be made to rise and fall by changing the voltage across the plates.

②The controversy. Some controversy was raised by historian Gerald Holton who pointed out that Millikan recorded more measurements in his journal than he included in his final results. Holton suggested these data points were omitted from the large set of oil drops measured in his experiments without apparent reason. This claim was disputed by Allan Franklin, a high energy physics experimentalist and philosopher of science at the University of Colorado. Franklin contended that Millikan's exclusions of data did not substantively affect his final value of e, but did reduce the statistical error around this estimate e. This enabled Millikan to claim that he had calculated e to better than one half of one percent. In fact, if Millikan had included all of the data he had thrown out, the standard error of the mean would have been within 2%. While this would still have resulted in Millikan having measured e better than anyone else at the time, the slightly larger uncertainty might have allowed more disagreement with his results within the physics community. While Franklin left his support for Millikan's measurement with the conclusion that concedes that Millikan may have performed "cosmetic surgery" on the data, David Goodstein investigated the original detailed notebooks kept by Millikan, concluding that Millikan plainly states here and in the reports that he included only drops that had undergone a "complete series of observations" and excluded no drops from this group of complete measurements. Reasons for a failure to generate a complete observation include annotations regarding the apparatus setup, oil drop production, and atmospheric effects which invalidated, in Millikan's opinion (borne out by the reduced error in this set), a given particular measurement.

In a commencement address given at the California Institute of Technology (Caltech) in 1974 (and reprinted in *Surely You're Joking, Mr. Feynman*! in 1985 as well as in *The Pleasure of Finding Things Out* in 1999), physicist Richard Feynman noted: "We have learned a lot from experience about how to handle some of the ways we fool ourselves. One example: Millikan measured the charge on an electron by an experiment with falling oil drops, and got an answer which we now know not to be quite right. It's a little bit off because he had the incorrect value for the viscosity of air. It's interesting to look at the history of measurements of the charge of an electron, after Millikan. If you plot them as a function of time, you find that one is a little bit bigger than Millikan's, and the next one's a little bit bigger than that, and the next one's a little bit bigger than that, until finally they settle down to a number which is higher."

Why didn't they discover the new number was higher right away? It's a thing that scientists are ashamed of — this history — because it's apparent that people did things like this: When they got a number that was too high above Millikan's, they thought something must be wrong — and they would look for and find a reason why something might be wrong. When they got a number close to Millikan's value they didn't look so hard. And so they eliminated the numbers that were too far off, and did other things like that...

(3) On Newton's decomposition of sunlight with a prism.

In 1666, Isaac Newton demonstrated that white light could be broken up into its composite colors by passing it through a prism, then using a second prism to reassemble them. He showed that white light is compound light or combined light(Fig. 3.4.4). Before Newton, most scientists believed that white was the fundamental color of light like black.

Fig. 3.4.4   Illustration of a dispersive prism separating white light into the colours of the spectrum, as discovered by Newton

Light changes speed as it moves from one medium to another (for example, from air into the glass of a prism). This speed change causes the light to be refracted and to enter the new medium at a different angle (Huygens principle). The degree of bending of the light's path depends on the angle that the incident beam of light makes with the surface, and on the ratio between the refractive indices of the two media (Snell's law). The refractive index of many materials (such as glass) varies with the wavelength or color of the light used, a phenomenon known as <u>dispersion</u>. This causes light of different colors to be refracted differently and to leave the prism at different angles, creating an effect similar to a rainbow. This can be used to separate a beam of white light into its constituent spectrum of colors. A similar separation happens with iridescent materials, such as a soap bubble. Prisms will generally disperse light over a much larger frequency bandwidth than diffraction gratings, making them useful for broad-spectrum spectroscopy. Furthermore, prisms do not suffer from complications arising from overlapping spectral orders, which all gratings have.

In fact, Newton observed that the spectrum of colours exiting a prism in the position of minimum deviation is oblong, even when the light ray entering the prism is circular, which is to say, the prism refracts different colours by different angles. This led him to conclude that colour is a property intrinsic to light — a point which had been debated in prior years.

Newton showed that coloured light does not change its properties by separating out a coloured beam and shining it on various objects, and that regardless of whether reflected, scattered, or transmitted, the light remains the same colour. Thus, he observed that colour is the result of objects interacting with already-coloured light rather than objects generating the colour themselves. This is known as Newton's theory of colour.

It should be pointed out that prisms are sometimes used for the internal reflection at the surfaces rather than for dispersion. If light inside the prism hits one of the surfaces at a sufficiently steep angle, total internal reflection occurs and all of the light is reflected. This makes a prism a useful substitute for a mirror in some situations.

White light can be generated by the sun, by stars, or by earthbound sources such as fluorescent lamps, white LEDs and incandescent bulbs. On the screen of a color television or computer, white is produced by mixing the primary colors of light: red, green and blue (RGB) at full intensity, a process called additive mixing (Fig. 3.4.5). White light can be fabricated using light with only two wavelengths, for instance by mixing light from a red and cyan laser or yellow and blue lasers. This light will however have very few practical applications since color rendering of objects will be greatly distorted.

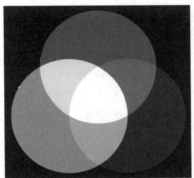

Fig. 3.4.5   In the RGB color model, used to create colors on TV and computer screens, white is made by mixing red, blue and green light at full intensity

The fact that light sources with vastly different spectral power distributions can result in a similar sensory experience is due to the way the light is processed by the visual system. One color that arises from two different spectral power distributions is called a metamerism.

Many of the light sources that emit white light emit light at almost all visible wavelengths (sun light, incandescent lamps of various color temperatures). This has led to the notion that white light can be defined as a mixture of "all colors" or "all visible wavelengths". This widespread idea is a misconception, and might originally stem from the fact that Newton discovered that sunlight is composed of light with wavelengths across the visible spectrum. Concluding that since "all colors" produce white light then white must be made up of "all colors" is a common logical error called affirming the consequent, which might be the cause of the misunderstanding.

A range of spectral distributions of light sources can be perceived as white — there is no single, unique specification of "white light". For example, when you buy a "white" light bulb, you might buy one labeled 2 700 K, 6 000 K, etc, which produce light having very different spectral distributions, and yet this will not prevent you from identifying the color of objects that they illuminate.

(4) On Young's light-interference experiment and its application in electron's double slits experiment.

Young's interference experiment, also called Young's double-slit interferometer, was the original version of the modern double-slit experiment, performed at the beginning of the nineteenth century by Thomas Young. This experiment played a major role in the general acceptance of the wave theory of light. In Young's own judgement, this was the most important of his many achievements.

During this period of 17th and 18th centuries, many scientists proposed a wave theory of light based on experimental observations, including Robert Hooke, Christiaan Huygens and Leonhard Euler. However, Isaac Newton, who did many experimental investigations of light, had rejected the wave theory of light and developed his corpuscular theory of light according to which light is emitted from a luminous body in the form of tiny particles. This theory held sway until the beginning of the nineteenth century despite the fact that many phenomena, including diffraction effects at edges or in narrow apertures, colours in thin films and insect wings, and the apparent failure of light particles to crash into one another when two light beams crossed, could not be adequately explained by the corpuscular theory which, nonetheless, had many eminent supporters, including Pierre-Simon Laplace and Jean-Baptiste Biot.

While studying medicine at Göttingen in the 1790s, Thomas Young wrote a thesis on the physical and mathematical properties of sound and in 1800, he presented a paper to the Royal Society (written in 1799) where he argued that light was also a wave motion. His idea was greeted with a certain amount of skepticism because it contradicted Newton's corpuscular theory.

Nonetheless, he continued to develop his ideas. He believed that a wave model could much better explain many aspects of light propagation than the corpuscular model.

A very extensive class of phenomena leads us still more directly to the same conclusion. They consist chiefly of the production of colours by means of transparent plates, and by diffraction or inflection, none of which have been explained upon the supposition of emanation, in a manner sufficiently minute or comprehensive to satisfy the most candid even of the advocates for the projectile system, while on the other hand, all of them may be at once understood, from the effect of the interference of double lights, in a manner nearly similar to that which constitutes in sound the sensation of a beat, when two strings forming an imperfect unison, are heard to vibrate together.

He demonstrated the phenomenon of interference in water waves (Fig. 3.4.6).

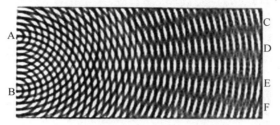

Fig. 3.4.6 Thomas Young's sketch of two-slit interference based on observations of water waves

In 1801, he presented a famous paper to the Royal Society entitled "On the Theory of Light and Colours" which described various interference phenomena, and in 1803 he described his famous double-slit experiment (Fig. 3.4.7, Fig. 3.4.8). Strictly speaking, there was no double slit in the original experiment as described by Young. Instead, the sunlight reflected off a steering mirror passed through a small hole in a paper, and the resulting thin light beam was then split in half alongside a paper card.

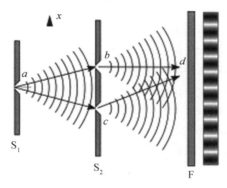

Fig. 3.4.7 Modern illustration of the experiment

Supposing the light of any given colour to consist of undulations of a given breadth, or of a given frequency, it follows that these undulations must be liable to those effects which we have already examined in the case of the waves of water and the pulses of sound. It has been shown that two equal series of waves, proceeding from centers near each other, may be seen to destroy each other's effects at certain points, and at other points to redouble them, and the beating of two sounds has been explained from a similar interference. We are now to apply the same principles to the alternate union and extinction of colours.

In order that the effects of two portions of light may be thus combined, it is necessary that they are derived from the same origin, and that they arrive at the same point by different paths, in directions not much deviating from each other. This deviation may be produced in one or both of the portions by diffraction, by reflection, by refraction, or by any of these effects combined. But the simplest case appears to be, when a beam of homogeneous light

Physics Laboratory Manual

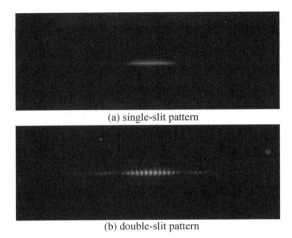

Fig. 3.4.8　Same double-slit assembly (0.7 mm between slits); in top image, one slit is closed. In the single-slit image, a diffraction pattern (the faint spots on either side of the main band) forms due to the nonzero width of the slit. A diffraction pattern is also seen in the double-slit image, but at twice the intensity and with the addition of many smaller interference fringes

falls on a screen in which there are two very small holes or slits, which may be considered as centers of divergence, from whence the light is diffracted in every direction. In this case, when the two newly formed beams are received on a surface placed so as to intercept them, their light is divided by dark stripes into portions nearly equal, but becoming wider as the surface is more remote from the apertures, so as to subtend very nearly equal angles from the apertures at all distances, and wider also in the same proportion as the apertures are closer to each other. The middle of the two portions is always light, and the bright stripes on each side are at such distances, that the light coming to them from one of the apertures, must have passed through a longer space than that which comes from the other, by an interval which is equal to the breadth of one, two, three, or more of the supposed undulations, while the intervening dark spaces correspond to a difference of half a supposed undulation, of one and a half, of two and a half, or more.

From a comparison of various experiments, it appears that the breadth of the undulations constituting the extreme red light must be supposed to be, in air, about one 36 thousandth of an inch, and those of the extreme violet about one 60 thousandth; the mean of the whole spectrum, with respect to the intensity of light, being about one 45 thousandth. From these dimensions it follows, calculating upon the known velocity of light, that almost 500 millions of millions of the slowest of such undulations must enter the eye in a single second. The combination of two portions of white or mixed light, when viewed at a great distance, exhibits a few white and black stripes, corresponding to this interval: although, upon closer inspection, the distinct effects of an infinite number of stripes of different breadths appear to be compounded together, so as to produce a beautiful diversity of tints, passing by degrees into each other. The central whiteness is first changed to a yellowish, and then to a tawny colour, succeeded by crimson, and by violet and blue, which together appear, when seen at a distance, as a dark stripe; after this a green light appears, and the dark space beyond it has a crimson hue; the subsequent lights are all more or less green, the dark spaces purple and reddish; and the red light appears so far to predominate in all these effects, that the red or purple stripes occupy nearly the same place in the mixed fringes as if their light were received separately.

In the years 1803 – 1804, a series of unsigned attacks on Young's theories appeared in the Edinburgh Review. The anonymous author (later revealed to be Henry Brougham, a founder of the Edinburgh Review) succeeded in undermining Young's credibility among the reading public sufficiently that a publisher who had committed to publishing Young's Royal Institution lectures backed out of the deal. This incident prompted Young to focus more on his medical practice and less on physics.

①Acceptance of the wave theory of light. In 1817, the corpuscular theorists at the French Academy of Sciences which included Siméon Denis Poisson were so confident that they set the subject for the next year's prize as diffraction, being certain that a particle theorist would win it. Augustin-Jean Fresnel submitted a thesis based on wave theory and whose substance consisted of a synthesis of the Huygens' principle and Young's principle of interference.

Poisson studied Fresnel's theory in detail and of course looked for a way to prove it wrong being a supporter of the particle theory of light. Poisson thought that he had found a flaw when he argued that a consequence of Fresnel's theory was that there would exist an on-axis bright spot in the shadow of a circular obstacle blocking apoint source of light, where there should be complete darkness according to the particle-theory of light. Fresnel's theory could not be true, Poisson declared: surely this result was absurd. (The Poisson spot is not easily observed in everyday situations, because most everyday sources of light are not good point sources.)

However, the head of the committee, Dominique-François-Jean Arago thought it was necessary to perform the experiment in more detail. He molded a 2-mm metallic disk to a glass plate with wax. To everyone's surprise, he succeeded in observing the predicted spot, which convinced most scientists of the wave-nature of light. In the end Fresnel won the competition.

After that, the corpuscular theory of light was vanquished, not to be heard of again till the 20th century. Arago later noted that the phenomenon (which was later to be known as the Arago spot) had already been observed by Joseph-Nicolas Delisle and Giacomo F. Maraldi a century earlier.

In modern physics, the double-slit experiment is a demonstration that light and matter can display characteristics of both classically defined waves and particles (wave-particle duality); moreover, it displays the fundamentally probabilistic nature of quantum mechanical phenomena. The experiment was first performed with light by Thomas Young in 1801. In 1927, Davisson and Germer demonstrated that electrons show the same behavior, which was later extended to atoms and molecules (Fig. 3.4.9).

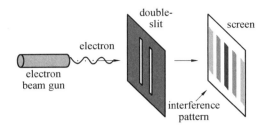

Fig. 3.4.9 Photons or particles of matter (like an electron) produce a wave pattern when two slits are used

Thomas Young's experiment with light was part of classical physics well before quantum mechanics, and the concept of wave-particle duality. He believed it demonstrated that the wave theory of light was correct, and his experiment is sometimes referred to as Young's experimenter Young's slits.

The experiment belongs to a general class of "double path" experiments, in which a wave is split into two separate waves that later combine into a single wave. Changes in the path lengths of both waves result in a phase shift, creating an interference pattern. Another version is the Mach – Zehnder interferometer, which splits the beam with a mirror. In the basic version of this experiment, a coherent light source, such as a laser beam, illuminates a plate pierced by two parallel slits, and the light passing through the slits is observed on a screen behind the plate. The wave nature of light causes the light waves passing through the two slits to interfere, producing bright and dark bands on the screen — a result that would not be expected if light consisted of classical particles. However, the light is always found to be absorbed at the screen at discrete points, as individual particles (not waves), the interference pattern appearing via the varying density of these particle hits on the screen. Furthermore, versions of the experiment that include detectors at the slits find that each detected photon passes through one slit (as would a classical particle), and not through both slits (as would a wave). However, such experiments demonstrate that particles do not form the interference pattern if one detects which slit they pass through. These results demonstrate the principle of wave-particle duality.

Other atomic-scale entities, such as electrons, are found to exhibit the same behavior when fired towards a double slit. Additionally, the detection of individual discrete impacts is observed to be inherently probabilistic, which is inexplicable using classical mechanics.

The experiment can be done with entities much larger than electrons and photons, although it becomes more difficult as size increases. The largest entities for which the double-slit experiment has been performed were molecules that each comprised 810 atoms (whose total mass was over 10 000 atomic mass units).

The double-slit experiment (and its variations) has become a classic thought experiment, for its clarity in expressing the central puzzles of quantum mechanics. Because it demonstrates the fundamental limitation of the ability of the observer to predict experimental results, Richard Feynman called it "a phenomenon which is impossible [...] to explain in any classical way, and which has in it the heart of quantum mechanics. In reality, it contains the only mystery [of quantum mechanics]."

If light consisted strictly of ordinary or classical particles, and these particles were fired in a straight line through a slit and allowed to strike a screen on the other side, we would expect to see a pattern corresponding to the size and shape of the slit. However, when this "single-slit experiment" is actually performed, the pattern on the screen is a diffraction pattern in which the light is spread out. The smaller the slit, the greater the angle of spread. The top portion of the image shows the central portion of the pattern formed when a red laser illuminates a slit and, if one looks carefully, two faint side bands. More bands can be seen with a more highly refined apparatus. Diffraction explains the pattern as being the result of the interference of light waves from the slit.

If one illuminates two parallel slits, the light from the two slits again interferes. Here the interference is a more pronounced pattern with a series of alternating light and dark bands. The width of the bands is a property of the frequency of the illuminating light. (See the bottom photograph to the right.) When Thomas Young (1773 - 1829) first demonstrated this phenomenon, it indicated that light consists of waves, as the distribution of brightness can be explained by the alternately additive and subtractive interference of wavefronts. Young's experiment, performed in the early 1800s, played a vital part in the acceptance of the wave theory of light, vanquishing the corpuscular theory of light proposed by Isaac Newton, which had been the accepted model of light propagation in the 17th and 18th centuries. However, the later discovery of the photoelectric effect demonstrated that under different circumstances, light can behave as if it is composed of discrete particles. These seemingly contradictory discoveries made it necessary to go beyond classical physics and take the quantum nature of light into account.

Feynman was fond of saying that all of quantum mechanics can be gleaned from carefully thinking through the facts.

(5) On Cavendish's torsion - bar experiment.

The Cavendish experiment, performed in 1797 - 1798 by British scientist Henry Cavendish, was the first experiment to measure the force of gravity between masses in the laboratory and the first to yield accurate values for the gravitational constant. Because of the unit conventions then in use, the gravitational constant does not appear explicitly in Cavendish's work. Instead, the result was originally expressed as the specific gravity of the Earth, or equivalently the mass of the Earth. His experiment gave the first accurate values for these geophysical constants.

The experiment was devised sometime before 1783 by geologist John Michell, who constructed a torsion balance apparatus(Fig. 3.4.10 - 3.4.12) for it. However, Michell died in 1793 without completing the work. After his death, the apparatus passed to Francis John Hyde Wollaston and then to Cavendish, who rebuilt the apparatus but kept close to Michell's original plan. Cavendish then carried out a series of measurements with the equipment and reported his results in the Philosophical Transactions of the Royal Society in 1798.

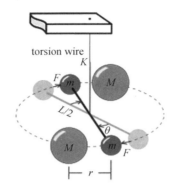

Fig. 3.4.10   Diagram of torsion balance

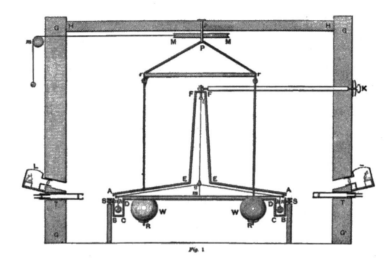

Fig. 3.4.11   Vertical section drawing of Cavendish's torsion balance instrument including the building in which it was housed. The large balls were hung from a frame so they could be rotated into position next to the small balls by a pulley from outside (from Figure 1 of Cavendish's paper)

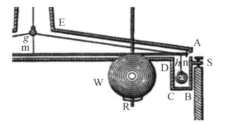

Fig. 3.4.12   Detail showing torsion balance arm (m), large ball (W), small ball (x), and isolating box (A,B,C,D,E)

The apparatus constructed by Cavendish was a torsion balance made of a six-foot (1.8 m) wooden rod horizontally suspended from a wire, with two 2-inch (51 mm) diameter 1.61-pound (0.73 kg) lead spheres, one attached to each end. Two 12-inch (300 mm)

348-pound (158 kg) lead balls were located near the smaller balls, about 9 inches (230 mm) away, and held in place with a separate suspension system. The experiment measured the faint gravitational attraction between the small balls and the larger ones.

The two large balls were positioned on alternate sides of the horizontal wooden arm of the balance. Their mutual attraction to the small balls caused the arm to rotate, twisting the wire supporting the arm. The arm stopped rotating when it reached an angle where the twisting force of the wire balanced the combined gravitational force of attraction between the large and small lead spheres. By measuring the angle of the rod and knowing the twisting force (torque) of the wire for a given angle, Cavendish was able to determine the force between the pairs of masses. Since the gravitational force of the Earth on the small ball could be measured directly by weighing it, the ratio of the two forces allowed the density of the Earth to be calculated, using Newton's law of gravitation.

Cavendish found that the Earth's density was (5.448 ± 0.033) times that of water (due to a simple arithmetic error, found in 1821 by Francis Baily, the erroneous value (5.480 ± 0.038) appears in his paper).

To find the wire's torsion coefficient, the torque exerted by the wire for a given angle of twist, Cavendish timed the natural oscillation period of the balance rod as it rotated slowly clockwise and counterclockwise against the twisting of the wire. The period was about 20 min. The torsion coefficient could be calculated from this and the mass and dimensions of the balance. Actually, the rod was never at rest; Cavendish had to measure the deflection angle of the rod while it was oscillating.

Cavendish's equipment was remarkably sensitive for its time. The force involved in twisting the torsion balance was very small, 1.74×10^{-7} N, about 1/50 000 000 of the weight of the small balls. To prevent air currents and temperature changes from interfering with the measurements, Cavendish placed the entire apparatus in a wooden box about 2 feet (0.61 m) thick, 10 feet (3.0 m) tall, and 10 feet (3.0 m) wide, all in a closed shed on his estate. Through two holes in the walls of the shed, Cavendish used telescopes to observe the movement of the torsion balance's horizontal rod. The motion of the rod was only about 0.16 in (4.1 mm). Cavendish was able to measure this small deflection to an accuracy of better than 0.01 in (0.25 mm) using vernier scales on the ends of the rod. The accuracy of Cavendish's result was not exceeded until C. V. Boys's experiment in 1895. In time, Michell's torsion balance became the dominant technique for measuring the gravitational constant (G) and most contemporary measurements still use variations of it.

Cavendish's result was also the first evidence for a planetary core made of metal. The result of 5.4 g $\cdot$ cm^{-3} is close to 80% of the density of liquid iron, and 80% higher than the density of the Earth's outer crust, suggesting the existence of a dense iron core.

The formulation of Newtonian gravity in terms of a gravitational constant did not become standard until long after Cavendish's time. Indeed, one of the first references to G is in 1873, 75 years after Cavendish's work.

Cavendish expressed his result in terms of the density of the Earth ρ_{earth}; he referred to his experiment in correspondence as "weighing the world". Later authors reformulated his results in modern terms:

$$G = g \frac{R_{earth}^2}{M_{earth}} = \frac{3g}{4\pi R_{earth} \rho_{earth}}$$

After converting to SI units, Cavendish's value for the Earth's density, 5.448 g · cm^{-3}, gives $G = 6.74 \times 10^{-11}$ m^3 · kg^{-1} · s^{-2}, which differs by only 1% from the 2014 CODATA value of $6.674\ 08 \times 10^{-11}$ m^3 · kg^{-1} · s^{-2}.

Physicists, however, often use units where the gravitational constant takes a different form. The Gaussian gravitational constant used in space dynamics is a defined constant and the Cavendish experiment can be considered as a measurement of this constant. In Cavendish's time, physicists used the same units for mass and weight, in effect taking g as a standard acceleration. Then, since R_{earth} was known, ρ_{earth} played the role of an inverse gravitational constant. The density of the Earth was hence a much sought-after quantity at the time, and there had been earlier attempts to measure it, such as the Schiehallion experiment in 1774.

For this reason, historians of science have argued that Cavendish did not measure the gravitational constant.

(6) On Eratosthenes' measurement of the Earth's circumference.

Earth's circumference is the distance around the Earth, either around the equator (40 075.017 km) or around the poles (40 007.863 km).

Measurement of Earth's circumference has been important to navigation since ancient times. It was first calculated by Eratosthenes, which he did by comparing altitudes of the mid-day sun at two places a known north-south distance apart (Fig. 3.4.13). In the Middle Ages, al-Biruni developed a way to perform the calculation from a single location and made a measurement at Nandana in present-day Pakistan.

In modern times, Earth's circumference has been used to define fundamental units of measurement of length: the nautical mile in the seventeenth century and the meter in the eighteenth. Earth's polar circumference is very near to 21 600 nautical miles because the nautical mile was intended to express 1/60 of a degree of latitude (i.e. 60 × 360), which is 21 600 partitions of the polar circumference. The polar circumference is even closer to 40 000 kilometers because the metre was originally defined to be one 10-millionth of the circumferential distance from pole to equator. The physical length of each unit of measure has remained close to what it was determined to be at the time, but the precision of measuring the circumference has improved since then.

Treated as asphere, determining Earth's circumference would be its single most important measurement (Earth actually deviates from a sphere by about 0.3% as characterized by flattening).

According to *Cleomedes' on the Circular Motions of the Celestial Bodies*, around

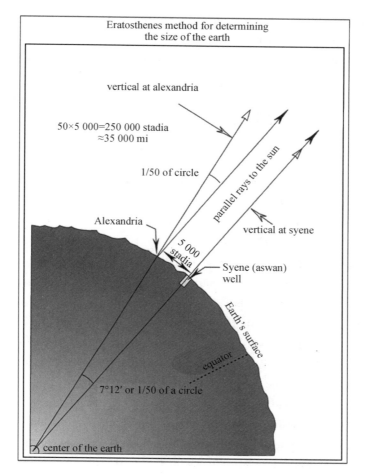

Fig. 3.4.13 Eratosthenes' method for determining the circumference of the Earth, with sunbeams shown as two rays hitting the ground at two locations in Egypt-Syene (Aswan) and Alexandria from outside

240 B.C., Eratosthenes, the librarian of the Library of Alexandria, calculated the circumference of the Earth in Ptolemaic Egypt. Using a scaphe, he knew that at local noon on the summer solstice in Syene (modern Aswan, Egypt), the Sun was directly overhead. (Syene is at latitude 24°05′ North, near to the Tropic of Cancer, which was 23°42′ North in 100 B.C..) He knew this because the shadow of someone looking down a deep well at that time in Syene blocked the reflection of the Sun on the water. He then measured the Sun's angle of elevation at noon in Alexandria by using a vertical rod, known as a gnomon, and measuring the length of its shadow on the ground. Using the length of the rod, and the length of the shadow, as the legs of a triangle, he calculated the angle of the sun's rays. This angle was about 7°, or 1/50 the circumference of a circle; taking the Earth as perfectly spherical, he concluded that the Earth's circumference was 50 times the known distance from Alexandria to Syene (5 000 stadia, a figure that was checked yearly), i.e. 250 000 stadia. Depending on whether he used the "Olympic stade" (176.4 m) or the Italian stade (184.8 m), this would

imply a circumference of 44 100 km (an error of 10%) or 46 100 km, an error of 15%. In 2012, Anthony Abreu Mora repeated Eratosthenes's calculation with more accurate data, the result was 40 074 km, which is 66 km different (0.16%) from the currently accepted polar circumference.

Posidonius calculated the Earth's circumference by reference to the position of the star Canopus. As explained by Cleomedes, Posidonius observed Canopus on but never above the horizon at Rhodes, while at Alexandria he saw it ascend as far as $7\frac{1}{2}$ degrees above the horizon (the meridian arc between the latitude of the two locales is actually 5 degrees 14 min). Since he thought Rhodes was 5 000 stadia due north of Alexandria, and the difference in the star's elevation indicated the distance between the two locales was 1/48 of the circle, he multiplied 5 000 by 48 to arrive at a figure of 240 000 stadia for the circumference of the earth. It is generally thought that the stadion used by Posidonius was almost exactly 1/10 of a modern statute mile. Thus Posidonius's measure of 240 000 stadia translates to 24 000 miles (39 000 km), not much short of the actual circumference of 24 901 miles (40 074 km). Strabo noted that the distance between Rhodes and Alexandria is 3 750 stadia, and reported Posidonius's estimate of the Earth's circumference to be 180 000 stadia or 18 000 miles (29 000 km). Pliny the Elder mentions Posidonius among his sources and without naming him reported his method for estimating the Earth's circumference. He noted, however, that Hipparchus had added some 26 000 stadia to Eratosthenes's estimate. The smaller value offered by Strabo and the different lengths of Greek and Roman stadia have created a persistent confusion around Posidonius's result. Ptolemy used Posidonius's lower value of 180 000 stades (about 33% too low) for the earth's circumference in his Geography. This was the number used by Christopher Columbus in order to underestimate the distance to India as 70 000 stades.

Around 830 A.D., Caliph Al-Ma'mun commissioned a group of Muslim astronomers led by Al-Khwarizmi to measure the distance from Tadmur (Palmyra) to Raqqa, in modern Syria. They calculated the Earth's circumference to be within 15% of the modern value, and possibly much closer. How accurate it actually was is not known because of uncertainty in the conversion between the medieval Arabic units and modern units, but in any case, technical limitations of the methods and tools would not permit a reliable measurement to better than about 5%.

A more convenient way to estimate was provided in Al-Biruni's Codex Masudicus. In contrast to his predecessors, who measured the Earth's circumference by sighting the Sun simultaneously from two different locations, Al-Biruni developed a new method of using trigonometric calculations, based on the angle between a plain and mountain top, which made it possible for it to be measured by a single person from a single location. From the top of the mountain, he sighted the dip angle which, along with the mountain's height (which he calculated beforehand), he applied to the law of sines formula. This was the earliest known use of dip angle and the earliest practical use of the law of sines. However, the method could not provide more accurate results than previous methods, due to technical limitations, and so Al-Biruni accepted the value calculated the previous century by the Al-Ma'mun expedition.

1 700 years after Eratosthenes's death, while Christopher Columbus studied what Eratosthenes had written about the size of the Earth, he chose to believe, based on a map by Toscanelli, that the Earth's circumference was 25% smaller. Had Columbus set sail knowing that Eratosthenes's larger circumference value was more accurate, he would have known that the place that he made landfall was not Asia, but rather the New World.

(7) On Galileo's experiments with rolling balls down inclined planes.

The experiment was described in *Galileo's Two New Sciences* published in 1638. One of Galileo's fictional characters, Salviati, describes an experiment using a bronze ball and a wooden ramp(Fig. 3.4.14). The wooden ramp was "12 cubits long, half a cubit wide and three finger-breadths thick" with a straight, smooth, polished groove. The groove was lined with "parchment, also smooth and polished as possible". And into this groove was placed "a hard, smooth and very round bronze ball". The ramp was inclined at various angles to slow the acceleration enough so that the elapsed time could be measured. The ball was allowed to roll a known distance down the ramp, and the time taken for the ball to move the known distance was measured. The time was measured using a water clock described as follows: "a large vessel of water placed in an elevated position; to the bottom of this vessel was soldered a pipe of small diameter giving a thin jet of water, which we collected in a small glass during the time of each descent, whether for the whole length of the channel or for a part of its length; the water thus collected was weighed, after each descent, on a very accurate balance; the differences and ratios of these weights gave us the differences and ratios of the times, and this with such accuracy that although the operation was repeated many, many times, there was no appreciable discrepancy in the results."

Fig. 3.4.14 Galileo's experiments with rolling balls down inclined planes (From physics – animation. com)

Galileo found that for an object in free fall, the distance that the object has fallen is always proportional to the square of the elapsed time:

$$\text{Distance} \propto \text{time}^2$$

Galileo had shown that objects in free fall under the influence of the Earth's gravitational field have a constant acceleration, and Galileo's contemporary, Johannes Kepler, had shown that the planets follow elliptical paths under the influence of the Sun's gravitational mass.

However, Galileo's free fall motions and Kepler's planetary motions remained distinct during Galileo's lifetime

According to legend, Galileo dropped balls of various densities from the Tower of Pisa and found that lighter and heavier ones fell at almost the same speed. His experiments actually took place using balls rolling down inclined planes, a form of falling sufficiently slow to be measured without advanced instruments.

In a relatively dense medium such as water, a heavier body falls faster than a lighter one. This led Aristotle to speculate that the rate of falling is proportional to the weight and inversely proportional to the density of the medium. From his experience with objects falling in water, he concluded that water is approximately ten times denser than air. By weighing a volume of compressed air, Galileo showed that this overestimates the density of air by a factor of forty. From his experiments with inclined planes, he concluded that if friction is neglected, all bodies fall at the same rate (which is also not true, since not only friction but also density of the medium relative to density of the bodies has to be negligible, aristotle correctly noticed that medium density is a factor but focused on body weight instead of density, Galileo neglected medium density which led him to correct conclusion for vacuum).

Galileo also advanced a theoretical argument to support his conclusion. He asked if two bodies of different weights and different rates of fall are tied by a string, does the combined system fall faster because it is now more massive? Or does the lighter body in its slower fall hold back the heavier body? The only convincing answer is neither: all the systems fall at the same rate.

Followers of Aristotle were aware that the motion of falling bodies was not uniform, but picked up speed with time. Since time is an abstract quantity, the peripatetics postulated that the speed was proportional to the distance. Galileo established experimentally that the speed is proportional to the time, but he also gave a theoretical argument that the speed could not possibly be proportional to the distance. In modern terms, if the rate of fall is proportional to the distance, the differential expression for the distance y travelled after time t is

$$\frac{dy}{dt} \propto y$$

with the condition that $y(0) = 0$. Galileo demonstrated that this system would stay at $y = 0$ for all time. If a perturbation set the system into motion somehow, the object would pick up speed exponentially in time, not quadratically.

Standing on the surface of the Moon in 1971, David Scott famously repeated Galileo's experiment by dropping a feather and a hammer from each hand at the same time. In the absence of a substantial atmosphere, the two objects fell and hit the Moon's surface at the same time.

The first convincing mathematical theory of gravity, in which two masses are attracted toward each other by a force whose effect decreases according to the inverse square of the distance between them — was Newton's law of universal gravitation. This, in turn, was replaced by the General theory of relativity due to Albert Einstein.

(8) On Rutherford's discovery of the nucleus.

The Rutherford model was devised by Ernest Rutherford to describe an atom. Rutherford directed the Geiger-Marsden experiment in 1909 which suggested, upon Rutherford's 1911 analysis, that J. J. Thomson's plum pudding model of the atom was incorrect. Rutherford's new model for the atom, based on the experimental results, contained new features of a relatively high central charge concentrated into a very small volume in comparison to the rest of the atom and with this central volume also containing the bulk of the atomic mass of the atom. This region would be known as the "nucleus" of the atom(Fig. 3.4.15).

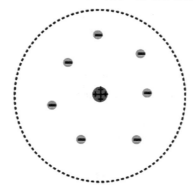

Fig. 3.4.15   Basic diagram of the atomic nuclear model: solid nucleus surrounded by the electrons

Rutherford overturned Thomson's model in 1911 with his well-known gold foil experiment in which he demonstrated that the atom has a tiny and heavy nucleus. Rutherford designed an experiment to use the alpha particles emitted by a radioactive element as probes to the unseen world of atomic structure. If Thomson was correct, the beam would go straight through the gold foil. In fact, most of the beams went through the foil, but a few were deflected(Fig. 3.4.16).

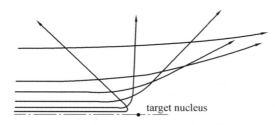

Fig. 3.4.16   The Rutherford scattering model

Rutherford presented his own physical model for subatomic structure, as an interpretation for the unexpected experimental results. In it, the atom is made up of a central charge (this is the modern atomic nucleus, though Rutherford did not use the term "nucleus" in his paper) surrounded by a cloud of (presumably) orbiting electrons. In this May 1911 paper, Rutherford only committed himself to a small central region of very high positive or negative charge in the atom.

For concreteness, consider the passage of a high speed α particle through an atom having a positive central charge Ne, and surrounded by a compensating charge of N electrons.

From purely energetic considerations of how far particles of known speed would be able to penetrate toward a central charge of 100 e, Rutherford was able to calculate that the radius of his gold central charge would need to be less (how much less could not be told) than 3.4×10^{-14} m. This was in a gold atom known to be 10^{-10} m or so in radius — a very surprising finding, as it implied a strong central charge less than 1/3 000 of the diameter of the atom.

The Rutherford model served to concentrate a great deal of the atom's charge and mass to a very small core, but didn't attribute any structure to the remaining electrons and remaining atomic mass. It did mention the atomic model of Hantaro Nagaoka, in which the electrons are arranged in one or more rings, with the specific metaphorical structure of the stable rings of Saturn. The plum pudding model of J. J. Thomson also had rings of orbiting electrons. Jean Baptiste Perrin claimed in his Nobel lecture that he was the first one to suggest the model in his paper dated 1901.

The Rutherford paper suggested that the central charge of an atom might be "proportional" to its atomic mass in Hydrogen mass units u (roughly 1/2 of it, in Rutherford's model). For gold, this mass number is 197 (not then known to great accuracy) and was therefore modeled by Rutherford to be possibly 196 u. However, Rutherford did not attempt to make the direct connection of central charge to atomic number, since gold's "atomic number" (at that time merely its place number in the periodic table) was 79, and Rutherford had modeled the charge to be about + 100 units (he had actually suggested 98 units of positive charge, to make half of 196). Thus, Rutherford did not formally suggest the two numbers (periodic table place, 79, and nuclear charge, 98 or 100) might be exactly the same.

A month after Rutherford's paper appeared, the proposal regarding the exact identity of atomic number and nuclear charge was made by Antonius van den Broek, and later confirmed experimentally within two years, by Henry Moseley.

The followings are the key indicators.

①The atom's electron cloud does not influence alpha particle scattering.

②Much of an atom's positive charge is concentrated in a relatively tiny volume at the center of the atom, known today as the nucleus. The magnitude of this charge is proportional to (up to a charge number that can be approximately half of) the atom's atomic mass — the remaining mass is now known to be mostly attributed to neutrons. This concentrated central mass and charge is responsible for deflecting both α and β particles.

③The mass of heavy atoms such as gold is mostly concentrated in the central charge region, since calculations show it is not deflected or moved by the high speed alpha particles, which have very high momentum in comparison to electrons, but not with regard to a heavy atom as a whole.

④The atom itself is about 100 000 (10^5) times the diameter of the nucleus. This could be related to putting a grain of sand in the middle of a football field.

After Rutherford's discovery, scientists started to realize that the atom is not ultimately a single particle, but is made up of far smaller subatomic particles. Subsequent research determined the exact atomic structure which led to Rutherford's gold foil experiment. Scientists eventually discovered that atoms have a positively charged nucleus (with an exact atomic number of charges) in the center, with a radius of about 1.2×10^{-15} m × (atomic mass number)$^{1/3}$. Electrons were found to be even smaller.

Later, scientists found the expected number of electrons (the same as the atomic number) in an atom by using X-rays. When an X-ray passes through an atom, some of it is scattered, while the rest passes through the atom. Since the X-ray loses its intensity primarily due to scattering at electrons, by noting the rate of decrease in X-ray intensity, the number of electrons contained in an atom can be accurately estimated.

(9) on Foucault's pendulum.

In 1851, Jean Bernard Léon Foucault showed that the plane of oscillation of a pendulum, like a gyroscope, tends to stay constant regardless of the motion of the pivot, and that this could be used to demonstrate the rotation of the Earth. He suspended a pendulum free to swing in two dimensions (later named the Foucault pendulum) from the dome of the Panthéon in Paris(Fig. 3.4.17). The length of the cord was 67 m (220 ft). Once the pendulum was set in motion, the plane of swing was observed to precess or rotate 360° clockwise in about 32 h. This was the first demonstration of the Earth's rotation that didn't depend on celestial observations, and a "pendulum mania" broke out, as Foucault pendulums were displayed in many cities and attracted large crowds.

The Foucault pendulum or Foucault's pendulum is a simple device named after French physicist Léon Foucault and conceived as an experiment to demonstrate the Earth's rotation. The pendulum was introduced in 1851 and was the first experiment to give simple, direct evidence of the Earth's rotation. Today, Foucault pendulums are popular displays in science museums and universities.

At either the North Pole or South Pole, the plane of oscillation of a pendulum remains fixed relative to the distant masses of the universe while Earth rotates underneath it, taking one sidereal day to complete a rotation. So, relative to Earth, the plane of oscillation of a pendulum at the North Pole (viewed from above) undergoes a full clockwise rotation during one day; a pendulum at the South Pole rotates counterclockwise(Fig. 3.4.18).

When a Foucault pendulum is suspended at the equator, the plane of oscillation remains fixed relative to Earth. At other latitudes, the plane of oscillation precesses relative to Earth, but more slowly than at the pole, the angular speed, ω (measured in clockwise degrees per sidereal day), is proportional to the sine of the latitude, φ:

$$\omega = 360° \sin \varphi \text{ day}^{-1}$$

where latitudes north and south of the equator are defined as positive and negative, respectively. For example, a Foucault pendulum at 30° south latitude, viewed from above by an earthbound observer, rotates counterclockwise 360° in two days.

Fig. 3.4.17   Foucault's pendulum in the Panthéon, Paris

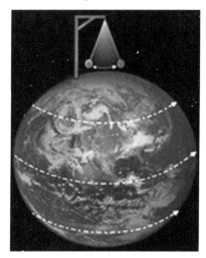

Fig. 3.4.18   Foucault pendulum at the North Pole: The pendulum swings
in the same plane as the Earth rotates beneath it

To demonstrate rotation directly rather than indirectly via the swinging pendulum, Foucault used a gyroscope in an 1852 experiment. The inner gimbal of the Foucault gyroscope was balanced on knife edge bearings on the outer gimbal and the outer gimbal was suspended by a fine, torsion-free thread in such a manner that the lower pivot point carried almost no weight. The gyro was spun to 9 000 – 12 000 revolutions per minute with an arrangement of gears before being placed into position, which was sufficient time to balance the gyroscope and carry out 10 minutes of experimentation. The instrument could be observed either with a microscope viewing a tenth of a degree scale or by a long pointer. At least three more copies of a Foucault gyro were made in convenient travelling and demonstration boxes and copies survive in the UK, France, and the US.

A Foucault pendulum requires care to set up because imprecise construction can cause additional veering which masks the terrestrial effect. The initial launch of the pendulum is critical; the traditional way to do this is to use a flame to burn through a thread which temporarily holds the bob in its starting position, thus avoiding unwanted sideways motion (see adetail of the launch at the 50th anniversary in 1902, Fig. 3.4.19).

Fig. 3.4.19   An excerpt from the illustrated supplement of the magazine Le Petit Parisien dated November 2, 1902, on the 50th anniversary of the experiment of Léon Foucault demonstrating the rotation of the earth

Air resistance damps the oscillation, so some Foucault pendulums in museums incorporate an electromagnetic or other drive to keep the bob swinging; others are restarted regularly, sometimes with a launching ceremony as an added attraction.

References

[1] YOUNG H D, FREEDMAN R A. University physics with modern physics [M]. 13th ed. San Francisco: Addison Wesley, 2012.

[2] HALLIDAY D, RESNICK R, WALKER J. Fundamentals of physics, extended edition [M]. 8th ed. Hoboken: Wiley, 2007.

[3] SERWAY R A, VUILLE C. College Physics [M]. 9th ed. Boston: Brooks Cole, 2011.

[4] FORNASINI P. The uncertainty in physical measurements: an introduction to data analysis in the physics laboratory [M]. New York: Springer, 2008.

[5] YAAKOV K. Experiments and demonstrations in physics [M]. London: World Scientific, 2007.

[6] WILSON J D, HERNÁNDEZ-HALL C A. Physics laboratory experiments [M]. 8th ed. Boston: Brooks Cole, 2014.

[7] LOYD D H. Physics laboratory manual [M]. 3rd ed. Belmont: Cengage Learning, 2007.

[8] HARIHARAN P. Basics of holography [M]. Cambridge: Cambridge University Press, 2002.

[9] BENNETT W R J R, MORRISON A C H. The science of musical sound [M]. Cham Switzerland: Springer, 2018.

[10] PARKER B. Good vibration: the physics of music [M]. Baltimore: Johns Hopkins University Press, 2009.